A PhD Is Not Enough!

A PhD IS
NOT ENOUGH!

A Guide to Survival in Science

REVISED
EDITION

PETER J. FEIBELMAN

BASIC BOOKS
A MEMBER OF THE PERSEUS BOOKS GROUP
New York

Books published by Basic Books are available at special
discounts for bulk purchases in the United States by
corporations, institutions, and other organizations. For
more information, please contact the Special Markets
Department at the Perseus Books Group, 2300 Chestnut
Street, Suite 200, Philadelphia, PA 19103, or call (800)
810–4145, ext. 5000, or e-mail
special.markets@perseusbooks.com.

Designed by Timm Bryson

Library of Congress Cataloging-in-Publication Data
Feibelman, Peter J.
   A PhD is not enough! : a guide to survival in science /
Peter J. Feibelman. — Rev. ed.
     p. cm.
   First published: Reading, Mass. : Addison-Wesley,
c1993.
   Includes bibliographical references.
   ISBN 978-0-465-02222-9 (alk. paper)
   1. Science—Vocational guidance—Handbooks,
manuals, etc. 2. Scientists—Training of—Handbooks,
manuals, etc. 3. Mentoring in the professions—
Handbooks, manuals, etc. I. Title.
   Q147.F45 2011
   502.3—dc22
                                    2010035289
Ebook ISBN: 978-0-465-02533-6
10 9 8 7 6 5 4

To Lori, Camilla, and Adam

CONTENTS

Contents

PREFACE: WHAT THIS BOOK IS ABOUT

My scientific career almost never happened. I emerged from graduate school with a PhD and excellent technical skills but with little understanding of how to survive in science. In this, I was not unusual. Survival skills are rarely part of the graduate curriculum. Many professional scientists believe that "good" students find their way on their own, while the remainder cannot be helped. This justifies neglect and, perhaps not incidentally, reduces work load. There may be some sense to the Darwinian selection process implicit in "benign neglect," but on the whole, failing to teach science survival results in wasting a great deal of student talent and time, and not infrequently makes a mess of students' lives.

Because science survival skills are rarely taught in a direct way, most young scientists need a mentor. Some will find one in graduate school, or as a postdoctoral researcher, or perhaps as an assistant professor. Those who do not have an excellent chance of moving from graduate study to scientific retirement without passing through a career. The unmentored can only succeed by being considerably more astute than the naive, idealistic, and very bright young persons who generally choose a science major.

These thoughts have been on my mind ever since I almost had to tell Mom and Dad that their golden boy was not good enough to find a permanent (or any!) job in physics, a job for which his qualifications included eight years of higher education and four more of postdoctoral work. The agony of those days is not easily forgotten—the boy with the high IQ, who had skipped a grade, graduated from the Bronx High School of Science at 16 and from Columbia summa cum laude at 20, found himself in a muddle at 28. How do you choose a research problem? How do you give a talk? What do you do to persuade a university or a national or industrial lab to hire and keep you? I hadn't a clue until, midway through my second postdoctoral job, I had the good fortune to spend some months collaborating with a young professor who cared whether I survived as a scientist. Although this mentoring relationship was brief, it helped me acquire a set of skills that graduate education did not, skills without which my lengthy training in physics would have been wasted.

This book is meant for those who will not be lucky enough to find a mentor early, for those who naively suppose that getting through graduate school, doing a postdoc, etc., are enough to guarantee a scientific career. I want you to see what stands between you and a career, to help you prepare for the inevitable obstacles before they overwhelm you. In short, I hope to enable you to use your exceptional brainpower in the way that you and those who put you through school have dreamed about.

I begin with some brief case histories. This may help to put your own early career in better perspective. At least I hope it will give you a feeling for how important mentoring can be.

Important or not, you are likely to wonder whether an elder who emerged into the scientific marketplace when times were flush, and advanced technology looked very different from today's, can possibly offer you useful advice. Chapter 2 argues that one can.

Succeeding chapters are arranged in parallel with a career trajectory. Please skip ahead to whichever may be relevant to your situation. Chapter 3 deals with choosing a thesis or a postdoctoral adviser. My choice of thesis adviser was based on two criteria: Who is the most eminent professor in the department? And whose students finish soonest? Was this intelligent, or did it represent a first mistake? Chapter 4 concerns oral presentation of your work. However brilliant your insights, they will be of little use if you cannot make them appear interesting to others. If no one pays attention, what difference does it make if your results are clever? There are of course Nobel prize–winners whose orations are Delphic, whose visuals look as though they were put together during a particularly turbulent flight, and so on. But you are not one of them yet, and if that is how your talks are prepared, you never will be either. There is more to Chapter 4, though, than advice on preparing appealing slides. It contains a range of important ideas on making your oral presentations effective.

In Chapter 5, you will find a discussion of paper writing. Through your scholarly articles, you can make yourself known nationally and internationally. This means that your reputation in science does not just depend on what your boss says about you but also on documentation that is readily available on the Internet. You should therefore view publishing as a means to attaining job security and take the task of writing compelling journal articles very seriously.

Chapter 6 is devoted to career choices, mainly the merits and defects of positions in academia and in government or industrial labs. The focus is on being reflective and rational rather than naive or romantic about key decisions in your scientific life. In Chapter

7, I discuss job interviews. There is more to an interview than wearing your Sunday best and having a firm handshake. Doing your homework and persuading your potential employers that you have a sense of direction are the most important issues. Incidentally, this is not a matter of deception—knowing who your colleagues will be and developing an idea of what you want to know, scientifically, are keys to having a productive career. There are also a few choice words in this chapter about negotiations, once you do get an offer. Negotiating for what you will need when your leverage is maximal can make a large difference to your happiness and to your success.

In Chapter 8, I discuss what—to many—is the bane of scientific life, namely, getting money. This used to be the exclusive headache of those in academia, but nowadays it is also a significant part of the lives of government and industrial scientists. I suggest that you view the preparation of a proposal as an important scientific exercise. Coming to see and being able to articulate how your work fits into "the big picture" is essential not only to winning financial support but also to being a first-class researcher. Learning to distinguish extravagant "pie in the sky" from promises that you have a chance of fulfilling is also very valuable.

The most difficult problem in being a scientist is selecting what to work on, and it is even more difficult when you are just launching your career. Therefore, in Chapter 9, I venture a few comments on establishing a research program. Jumping into the hottest research area may not be a very good idea, nor is taking on a project that you have no realistic hope of completing before your short-term employment comes to an end. The main idea is to establish a program that simultaneously maximizes your chances of continuing employment *and* of scientific achievement. The focus is on strategic thinking.

As this book is written, economic times are tough worldwide, and funding for scientific research is contracting. I hardly need to emphasize that when resources become scarce, competition intensifies for what remains available. To win a permanent position in scientific research, and the funds to carry on serious work, you will have to be exceptionally thoughtful about your career choices. My hope is that this "pocket mentor" will help you to become more introspective about what it will take to succeed.

—ALBUQUERQUE, NM
August 1993 (updated in January 2010)

The past seventeen years have seen revolutionary changes in how we communicate information. Virtually all journals are available electronically. Preprints can be published on the Internet before or without ever being refereed. Overhead projectors have disappeared from scientific meetings in favor of LCD projectors and laptop computers. Résumés are often distributed electronically. This update of *A PhD Is Not Enough!* comes abreast of these changes, though the basic content of the 1993 original remains timely. The communications revolution cannot be ignored but has not made it less important to be thoughtful about choosing your career path or to respect audiences and readers. I still attend talks that make me squirm and struggle to read sleep-inducing scientific articles. I hope attentive readers of this book will reap the rewards of doing better.

—ALBUQUERQUE, NM
January 2010

ACKNOWLEDGMENTS

To make this handbook accessible to people whose backgrounds, experiences, and scientific interests differ from my own, I have prevailed on several friends and colleagues for advice. I am very grateful to Professors Michael J. Weber and Alison P. Weber of the University of Virginia for numerous constructive criticisms of the first draft. I also thank Dr. Ellen Stechel, my colleague at Sandia National Laboratories, and Professor George Luger of the University of New Mexico for their critical readings of the manuscript. Lastly, I thank my wife, Lori, for many editorial improvements.

CHAPTER 1

Do You See Yourself in This Picture?

The brief stories in this chapter have a common theme: that understanding and dealing rationally with the realities of a life in science are as important to science survival as being bright. Once you leave graduate school, the clock is ticking. Unlike a fine wine, you do not have many years to mature. As a young professional, you must be able to select appropriate research problems, you have to finish projects in a timely manner, and you ought to be giving compelling talks and publishing noteworthy papers. When job opportunities present themselves, you should be able to assess their value realistically. Romanticizing your prospects

is a major mistake and is likely to have serious consequences, not excluding dropping out of scientific life
prematurely. The first story is an excerpt from my own
scientific beginnings. The others are also nonfiction,
though I have altered locations and personal charac-
teristics to avoid invading the privacy of the protago-
nists. I have deliberately identified the various
characters with initials, rather than names, to avoid any
ethnic implications.

What Do Scientists Do?
Technique Versus Problem Orientation

Virtually all classroom work and much of what happens
in a typical thesis project is aimed at developing a student's technical skills. But although the success of your
research efforts may depend heavily on designing a
piece of apparatus or a computer code, and on making
it work properly, *no technical skill is worth more than
knowing how to select exciting research projects.* Regret-
tably, this vital ability is almost never taught. When I
signed on with a research adviser in my first year of
graduate school, I was thrilled to be given a problem
to work in the physics of the upper atmosphere. That
I had no idea what motivated the problem did not prevent me from carrying out an analysis, on a supercom-
puter of the day, and publishing my first paper at the

age of 22. For my thesis, I consciously switched to a project that would require learning the tools of modern quantum physics, but again I found myself assimilating technical skills without ever grasping the significance of the problem, without understanding how or whether it was at the cutting edge of science. This way of working became a habit, one that seriously threatened my career. My first seven publications were in seven different areas of physics. In each case, I relied on a senior scientist to tell me what would be an interesting problem to work on; then I would carry out the task. I assume it was my ability to complete projects that impressed my superiors sufficiently to keep me employed. It certainly wasn't my depth in any field.

Four years and two postdoctoral positions after earning a PhD—still having little sense of what I wanted to learn as a scientist—I was on the job market. More than anything else, I needed good recommendations from faculty at the university where I was employed. I was asked to give the weekly solid-state physics seminar and realized, at best dimly, that my performance in this venue was either going to make or break me as a scientist.

The talks I was giving at this point in my career reflected my approach to science. There was little in the way of introductory material. Much of the presentation was technical. I would describe a few "interesting" problems I had worked on and explain the methods I had used but would give little idea of context because I really didn't know what it was. For the seminar at hand, I prepared my usual hodgepodge of this project and that, with no introduction, no theme, and ultimately no meaning to anyone but an expert. Fortunately, the professor supervising my research, C., understood what was about to happen to me, and asked for a preview of my seminar in his office. Thank goodness I accepted this invitation. C. expressed surprise at how poorly I had prepared my talk (though I don't think he was surprised at all), how little grasp I seemed to have of the reasons that the problems we had worked out were meaningful, and consequently how uninterestingly I was going to present them to my audience. But, he told me, he thought I was too good technically to be allowed to fail in the way I was about to, and he gave me the lesson I needed.

His most important advice was:

1. There has to be a theme to your work—some objective—something you want to know. There has to be a story line. (Do not start with, "I have been trying to explain the interesting wavelength dependence of light scattering from small particles," but rather "There is a widespread need to explain to one's kids why the sky is blue.")

2. If you know why you have chosen to work on a particular problem, it is easy to present an absorbing seminar. Start out by telling your story, why the field you are working in is an important one, and what the main problems are. Give some historical material showing where the field is, the relative advantages of different methods, and so on. Then outline what you did, and describe your results. Conclude with a statement of how your results have advanced our understanding of nature, and perhaps give an inkling of the new directions that your work opens up. Do not assume that your audience comprises experts only. There may be a couple of them, but even experts like to hear things that they understand and particularly to have their colleagues hear (from someone else) why their field is an important one.

3. Lastly, rehearse your talk in front of one or two of your peers or professional supporters. Choose listeners who will not be shy about asking questions and offering constructive suggestions. Giving a seminar is serious business. Your future depends on the strong recommendations of your senior colleagues. If your talk is a hodgepodge of techniques or experiments or equations, if you seem to have no idea where you are headed, if you reek of deference to the experts in the audience, you

will not be perceived as a rising star, a budding scientific leader. You will fail.

The wonderful result of C.'s mentoring was that I finally learned what it means to be a scientist. In making my work meaningful to others, I had also made it compelling to myself. No longer was I just working on somebody else's problems. I was part of an intellectual enterprise with relatively well-defined goals, which might actually make a difference to humanity. I scrapped most of the equations I had planned to show and refocused my talk using thematic material I had garnered from C. I gave an excellent seminar—people I scarcely knew complimented me afterward on my choice of an exciting research area and remarked on the clarity of my presentation. In science, the reinforcement doesn't get much more positive than that. I had learned a key lesson and was on my way.

Timing Is Everything

Having completed a respectable thesis problem and having acquired a reputation in graduate school as an excellent sounding board and scientific consultant, T. accepted a postdoctoral position with a leading scientist at a first-rate government laboratory. There, he was offered and began to work on a computational research project that first involved arriving at a numerically practical mathematical formulation of a problem and then required a considerable computer programming effort. As the months passed, and with the necessity on the horizon of finding a permanent job, T. absorbed himself totally in his very challenging work. Whereas in graduate school, under little time pressure, he would have spent a few hours each week visiting labs and contributing to projects other than his own, as a postdoc, T. became utterly single-minded.

Working 12 hours a day and more, he managed to complete his computer program soon enough to be able to run test calculations. The results were promising but not far enough along to yield a persuasive "story." Accordingly, neither T. nor his audiences found his job seminar very exciting. What is more, since he had not taken time to meet and consult with scientists at his lab, his only strong recommendation was from his postdoctoral adviser. The lab itself was unwilling to promote T. to a permanent position, which it sometimes did, because he had not made himself useful, or even known, to a spectrum of its staff members.

On the outside, his job offers were a cut below what his thesis adviser had expected for him. In the competition for the best positions, T. did not persuade potential employers that he would ever derive useful results from his postdoctoral project, even though T. believed that he would have them within six months to a year. Other job candidates whose postdoctoral work had been far less ambitious, but had resulted in two or three finished projects, appeared much more impressive. Moreover, they had obtained excellent recommendations from the experimental colleagues whose data they had analyzed.

On the whole, it is hard to blame potential employers for their view of T. To them he was "a pig in a poke," an unknown quantity. His thesis work might just have been done by his thesis adviser, and his postdoctoral project, though in principle a worthy one, was unfinished. Would T. be able to complete projects on his own? Was he a self-starter? The information simply was not there, in the eyes of the interviewers.

To some extent, T.'s fate was the fault of his adviser. Assigning a long-term project to a postdoctoral researcher who will be on the job market in 18 months is a clear risk to the postdoc's future. But, had T. been as reflective about his career as he was in carrying out his research, he himself would have realized the dangerous path he was taking. As exciting as his assigned project seemed, he would have recognized that his postdoctoral years were the wrong time for such a large effort. At the very least, he would have reserved time each day or week to establish contact with other researchers at the lab and involved himself in one or two short-term projects with a clear chance for success. Many a graduate student or postdoc spends time trying to understand what his adviser wants and getting it done. In fact, it is the young scientists who define and carry out what *they* want, who learn to be scientific leaders, who find the best jobs and have the most productive and satisfying careers. Making your thesis or postdoctoral adviser happy is sensible, and worth doing, but not more so than acting in your own best interests.

Know Thyself—A Sweet Job Turns Sour

B. obtained a PhD from a top-flight university in the Midwest. He had two different thesis advisers during the course of his four years as a graduate student. The first was a Nobel prizewinner, a theoretician whose name is a household word to chemists. The second was an experimentalist, also a very widely respected scientist. Having completed his degree, and cognizant of the scarcity of real jobs, B. accepted a "permanent" position at a major laboratory instead of a postdoctoral, temporary slot. It did not take him long to realize that this apparently wonderful opportunity was a trap. On arrival at his new location, B. was presented with two options. A senior staff member, who was involved in a major experiment, suggested that B. begin his tenure by working in his lab. That way, B.'s knowledge of the experimental aspects of his field would deepen, and after a couple of years, he would be much better prepared to work on his own. Objectively, one would say that this was a wonderful opportunity, effectively a postdoctoral job, but at a regular staff salary and with a reasonable approximation to regular staff job security. B.'s alternative option was to begin independent work immediately. Talking to his younger colleagues, he heard that, in the eyes of management, a full staff member was supposed to run his own research program and that at the annual performance review, if he was perceived to be working as someone else's "assistant," his rating, salary, and job security would suffer, perhaps irretrievably.

One does not have to be a rocket scientist, as they say, to appreciate that B.'s two-year stint as a graduate student in experimental physics was inadequate preparation for him to perform at the level of his supposed peers. Nevertheless, unmentored, B. was not willing to risk his all-too-sweet regular staff position by choosing the training that he badly needed. This was a mistake. After three years of buying equipment and setting up a lab, B. had still not established a research program, and indeed had little idea of what he wanted to accomplish as a scientist. Thus, despite its investment in his laboratory equipment, and despite his nominally very impressive pedigree, B.'s employer moved him out of basic research. In an environment where goals were clearly defined from above, he eventually matured into a real contributor and is reasonably happy. On the other hand, he is not doing basic research any more, and he went through several very stressful years as a result of his bad start. Sadly, his failure at work coincided with the breakup of his marriage, an unhappy fate shared by many whose scientific careers flounder.

The PhD Technician

L. spent two postdoctoral years at a prestigious lab, switching into a new field. He had been hired as a postdoc there because of the technical know-how he had acquired as a graduate student. As a postdoctoral scientist, his task was to build a piece of equipment combining technology in his new area with that of his thesis work. The lab where he did his stint as a postdoc was satisfied enough with him. At the end of his two years, the desired instrument was in place, and L. had his name on a couple of publications with his postdoctoral adviser. Of course, it was recognized that L. had not really learned the basics of his new field, and so his postdoctoral employer did not offer him a permanent position.

A more aggressive or aware young man might have spent a significant fraction of his two years not simply building the desired instrument but also asking questions about the direction of his new field, reading as widely as possible in its literature, and formulating a research direction of his own. L. did not, however, and even at the end of his postdoc, no one had told him, nor did he realize that becoming an expert in a field and having an exciting research program is an essential aspect of being a scientist. L. did manage to land a "permanent" job after his postdoc. But as in B.'s case, permanency was an illusion.

In his new job, L. again built an instrument. But he never participated as an equal member in the group that hired him. At seminars or in planning research proposals, he had little to contribute. When he went before his manager to explain what his research plans were, he could say no more than that he planned to look around for "interesting" problems. L.'s employer was happy to possess the new instrument that he had built and got running. But it was not long before L. was moved from the research division of his company.

Some will argue that L. just wasn't suited for research, that his fate was predetermined by his personality. This may be the truth. On the other hand, I have the lingering feeling that if L. had been appropriately mentored at some point during his decade of higher education and as a postdoctoral researcher, he would have succeeded in the career for which he had trained, or perhaps would have switched earlier to a more appropriate field of specialization. It remains to be seen how well he will perform in his new job.

Institutionalized Conflict

Managers make many mistakes. More often than not, these hurt the people they manage rather than themselves. Consider F.'s experience as a postdoc in R.'s lab. R. had been hired after a two-year postdoctoral position but had the wit to appreciate that his "permanent" position would only *really* be permanent if he proved himself a capable scientist in his first two or three years. He invested his first year building a lab around a major piece of equipment and was ready to begin to do science when F. appeared at his threshold. F. had been hired to work on a project that seemed rather exciting to its managerial proponents but had failed to get the hoped-for, and necessary, external funding. The result was that management had to find something else for F. to do and had decided that because his training was similar to R.'s, F. would be a postdoc in R.'s lab. The results were inevitable. Being a clever young man, F. realized that his future depended on gaining recognition for a significant piece of work, work that would have to be done in short order. R., no less clever, understood that his probationary position required him to complete several projects and get credit for them. The result was not a happy collaboration but months of bickering over who would turn knobs on the machine and who would get credit for the scientific progress. Despite its responsibility for a bad situation, management did not like to hear the resultant whining from either side. F. ultimately won credit for most of the work done in R.'s lab, with the result that R., whose competence was felt to be more technical than scientific, was moved out of research. But management's distaste for F.'s complaining far exceeded its pleasure in his scientific achievements. F. was not considered as a candidate to replace the hapless R. He did eventually find another position in science, though, and I hope he will succeed.

Postmortem: Successful collaboration is possible when one or both contributors have established reputations, or when each researcher brings a different, identifiable skill to the collaborative project—for example, when a theorist and an experimentalist work together. Collaboration does not work, as a rule, for two young competitors. Neither F. nor R. was mature enough to realize that F.'s postdoc was a predictable nightmare, an arrangement that should have been rejected by both of them.

If F. and R. had found or had been assigned appropriate mentors early on, they might have been able to deal with the competitive relationship imposed on them. If management had explained to F. at the outset that R. was to be "the boss," and had discussed with both how credit for results was to be allocated, then F. could have made an informed decision on whether to work in R.'s lab, and he would have had little reason to complain later. However, on their own, F. and R. spent a miserable year and a half together, and R.'s scientific career is just a memory.

Impressing Mom and Dad: Whose Life Is It Anyway?

A common theme in the minds of young scientists is impressing Mom and Dad. This strong motivation is to be cherished, of course, but only if it does not overwhelm one's ability to make rational decisions. H. is the eldest daughter of a successful professor of microbiology. Having obtained a PhD in an area of limited interest to employers, she decided to switch fields, hoping her technical expertise would enable her to establish a niche. However, she decided to carry out this (wise) move as an assistant professor at a prestigious university (a questionable choice, at best).

A major factor in this decision was that she wanted to show her father that she could succeed in the academic world, just as he had. Had she thought her choice through, H. would have realized that when her dad was starting out, research funding was expanding dramatically, making the odds of success much better. She might also have foreseen that her next five years were going to be a major struggle, a period when any desires for a personal life would have to be put off. Between coming up to speed in her new field, fulfilling her teaching assignments, writing proposals, and building equipment—all essential before any research results could be produced—H. found herself spending 16-hour days in her office, the classroom, and her lab. Yes, she did receive tenure after five years. So in that sense she succeeded. But during those years, she had no life beyond her work, and by the time she was done, her marriage had disintegrated. Did this impress Dad?

In a national or industrial lab, H.'s plan would have been much easier to realize. With no teaching assignments, no committee meetings, no insistent students at the door wanting their grades explained, she could have made her name working eight or maybe ten hours per day. After five years of building a lab and producing science, she would have had little difficulty landing a tenured job at an excellent university. Meanwhile, she would have had time for her family—maybe even time to have the child she wanted. She would have been earning 30 to 60 percent more and would have had better job security. She might have relaxed with a good novel occasionally, or even taken a vacation. Things are working out for H. now, but she paid what I see as a high price for the romantic notion that she needed to move directly into academia to win her dad's approval.

Get a Mentor

I certainly hope that reading this book will help you recognize what is in your own interest. But no author can be expected to foresee your own special pitfalls. The best preparation you can make toward the goal of having a scientific career is to find yourself a "research aunt or uncle," someone with little or no authority over you, who has enough experience to act as a sounding board and to give accurate advice. Do not be shy about getting to know people outside your adviser's realm. The scientists at your lab will very likely cherish the human contact. They spend a lot of time behind the closed doors of lab and office, and everybody likes to give advice.

CHAPTER 2

Advice from a Dinosaur?

By my standards, today's world is technologically highly evolved. Very highly! With email in its infancy only a couple of decades ago, and long-distance phone calls costly, we dinosaurs mainly communicated by what is now disparaged as snail mail, if not in person. There was no Internet. Putting your résumé on a compact disc was not an option—indeed, to "burn a disc" had not entered the lexicon. A serious literature search involved many mind-numbing hours in a library (I know: "*What's a library?*"). Computers were unimaginably slow.

It is not just technology that has changed. Until the latter 1960s, for instance, widely held memories of the successful Manhattan Project, and worries spawned by the Soviet launch of *Sputnik* (in 1957), supported many, many dollars for physics and for science more broadly. Landing a tenure-track job, and even winning tenure itself, was not an especially taxing project for the fresh science PhD of that blessed era.

This perspective begs a serious question: Can you expect to find an effective mentor among scientists who succeeded in the technological and historical climate of two to four decades ago? The answer is yes, I contend, provided you narrow your search from those who are merely older to congenial researchers whose success has not clouded their historical and personal outlook. Notwithstanding an utter lack of interest in maintaining a Facebook page, a scientific elder can offer help in establishing a personal network of scientific contacts, in distinguishing an exciting research idea from a pedestrian one, in critiquing your oral and written expression, and so forth. That an elder researcher's path to tenure was relatively easy need not translate into his or her inability to distinguish good luck in emerging into the job market at a particularly blessed moment from having possessed superlative intellectual capacity and a clever career strategy.

Need I say that there are also plenty of scientific elders whose experience was not so different from your own, and who don't have to make a special effort to understand what you are facing? I am one of them. Despite receiving a PhD when times were still good, in December 1967, I made the "mistake" of accepting a postdoctoral position in Paris instead of immediately looking for a tenure-track job. I had a wonderful stay in France, but at the cost of then having to find a permanent research job in the hard times of the early 1970s instead of the easy ones of just a couple of years before. Not a seer, I had managed to place myself on the wrong side of a cusp in funding levels—and the right side for gaining an understanding of what a starting scientist must do in a tough economic environment to win a permanent place in the research community.

So, how did my quest come to a happy conclusion? In 1973, the U.S. economy was headed steeply downward; the Vietnam War was working toward its end ("not with a bang but a whimper"); the Watergate scandal was just months from forcing Richard Nixon to resign the presidency; and I, at age 31, was looking for a permanent job in physics. After two-plus years as a soft-money assistant professor, I'd been informed that when the three-year National Science Foundation grant that paid my salary expired, funds would not be available to move me to the tenure track. (Does this sound at all familiar?)

There were not many suitable jobs. I recall a trip to Texas to interview at the University of Houston, Texas

A&M, and UT–Austin. At each stop, I gave my talk, met privately with staff, felt I had done well, and was then informed that the position in question had evaporated. "Sorry about that!" In December, I spent five weeks on a research visit to the Stanford Applied Physics Department. One Sunday in Palo Alto, I noticed a job ad in the newspaper* for a scientist who would be hired to advise a mayor on the likely impact of urban development plans. The position was once again based on a finite-term grant. But, after two, two-year postdoctoral positions and a three-year assistant professorship, I was inured to the nomadic life, and so I applied. Despite my lack of credentials in urban planning, my interview, high up in San Francisco's stunning Transamerica building, went rather well, I thought, until I was asked, "What would you do if, a few weeks from now, you were offered a job in physics? Would you take it?" I gave an honest answer— the wrong answer, namely, "Yes." End of interview— back to despair.

But then, a bolt from the blue—a former postdoctoral colleague who had moved to Sandia Laboratories in New Mexico decided to quit research and become a medical doctor. He proposed my name as someone to fill his position, a permanent one. By then, I knew

---

* That's right—newspaper. Craigslist, need I remind you, did not exist in 1973.

what it takes to have a career in science. I could artic- ulate my research direction. I understood that as a theorist, I needed to persuade experimenters that I would be helpful to them, and also that I grasped ideas they did not. So, I prepared and burnished a talk. The first two-thirds of it were introductory, pictorial, and conceptual—deliberately designed to appeal to my hoped-for experimental colleagues. The last third was heavily theoretical, with equations, even, aimed at per- suading listeners that in me, they would be buying ex- pertise they themselves lacked.

These tactics worked! I was offered, and with alacrity accepted, a position at Sandia. On arrival, I did my ut- most to fulfill the promises I'd made in my interview— and as of the year 2011, at age 68, I've been a research scientist there for a very rewarding 36 years.

What lessons reside in this autobiographical extract and happy ending? One, not much of a surprise in a Facebook era, is that networking is an excellent way to gain opportunities. Responding to job ads may have the desired effect. Knowing someone is better.

Another lesson is the importance of being serious. Why would a hiring officer consider an applicant for an urban planning job who at the drop of a hat is pre- pared to return to the physics career he really wants? I wouldn't.

A third notion is that even in a market where few positions are available, the number is unlikely to be zero—and it is the best-prepared applicant who will win the competition. Having a reasonably good idea of what my Sandia interviewers would be hoping for, I spent serious time developing an appealing job talk. This was far from wasted effort.

Understand that the probability of landing a permanent job is the product of two factors. One is how many suitable positions are available. The other is your probability per job of being the successful candidate. There is essentially nothing you can do to affect the first factor. (Well . . . you might write your senator. Good luck with that!) Accordingly, it is a focus on the second factor that makes sense. Despairing over the unavailability of jobs wins you nothing. Preparing for an opportunity might—and in large measure, that is what this book is about. Its basic themes are:

1. Know thyself!
2. Understand and respect the needs of your audience.

Since my personal saga of 1973–1974, the U.S. and world economies have seen good times and bad. Twenty years on, with the United States once again in recession, the first printing of this book found a receptive readership. After the subsequent Internet boom came the Internet bust, and today we are experiencing and—only maybe—slowly emerging from the "Great

Recession" of 2008–2009. Once again, job opportunities for freshly minted scientists are scarce, and, accordingly, I am guessing you will find the advice from this dinosaur relevant, even in a world that, since 1993, has outwardly changed greatly.

CHAPTER 3

Important Choices
A Thesis Adviser, a Postdoctoral Job

As a young graduate student, I selected a thesis adviser on the bases of his prominence in the world of physics and his reputation as one who would not require me to spend too much time in graduate school. As with other aspects of my early career, I now see these criteria as reasonable but insufficient.

A Prominent Scientist as a Thesis Adviser

Choosing a prominent thesis adviser makes a lot of sense, but not because brilliance is transferable. It is not, as I have witnessed more than once. Trying to be another Linus Pauling, Roald Hoffmann, James Watson, or P. W. Anderson is a common road to failure. What a prominent adviser *can* offer is: 1. being part of the "old-boy network" (he or she can help you survive if times are tough, sometimes even if you don't deserve to); and 2. not competing with you. Point 1 is self-evident upon a moment's thought. Point 2 is not so obvious to the naive.

A young adviser, only recently on the road to a permanent research position, has a lot to prove, is understandably leery of being shown up by a student or postdoc, and is correspondingly unlikely to be generous with credit for ideas or progress. By contrast, advisers who have already made their mark view the accomplishments of their students, in effect their research "children," with pride, even joy. Thus, other things being equal, an established (tenured) professor is a superior choice for an adviser. This recommendation is a simple corollary of the way universities are organized. It is not an indictment of young professors to recognize that they are likely to view their own scientific survival as more important than that of their students.

A more senior adviser also offers you better prospects of finishing the thesis project that you start and of spending your entire graduate career at one university. Many assistant professors fail to win promotion to tenure. If this happens to your adviser, he or she will either have to move to another university or may drop out of academic science entirely. In either case, you will face unwanted, difficult choices: whether or not to move with your adviser, or whom to choose as a new one; whether to select a new dissertation topic or to try to find another professor who is willing and able to help you proceed in your initial direction.

Although a senior professor may also move to another job while you are a student, the probability is lower. One reason is that the bother involved in moving an established, large group is substantial. Another is that universities will offer what it takes, if the money is available, to retain their top staff. If your senior professorial adviser does decide to move, the consequences for your thesis project are unlikely to be dire. A senior scientist relocates by choice, usually because the funding situation in the new location is, or perhaps other aspects of scientific life are, better. Moving with your adviser is thus likely to be both financially possible and scientifically desirable. And if you do decide to move, the delay in your progress toward a PhD should be minimal.

Obviously, an older professor has a better chance of becoming seriously ill or dying while you are a student. Otherwise, the chances of a senior scientist's dropping out of research entirely are rather remote.

Tenure and prominence are not enough: Although signing on as the student of an established scientist has many clear advantages, choosing a good adviser is not as easy as finding out who has won the most important prizes, gives the most invited talks, or brings in the largest research grants. Is the professor you are considering available to consult with students on a reasonably frequent basis and able to convey real guidance? Is your intended adviser comfortable talking to people who are not scientific peers (i.e., beginners such as yourself)? Does the group you wish to join have a sense of purpose? Do its members interact with each other? And does Professor Eminent teach survival skills? These are important questions. Making a mistake in choosing your adviser can mean years of frustration. If you can learn the answers to the important questions in advance, by talking to current or former students, you may save yourself a lot of grief.

Do group members see the big picture? Prof. E. was obsessive. He was obnoxious. I have heard it said that he didn't know quantum mechanics. But his contributions to materials science were manifold—and his students have done wonderfully well. They knew what they wanted to learn, and they learned from each other. Thus, even if E. was often away consulting at industrial labs, his students thrived.

How do you find out in advance whether the group you are considering will be like E.'s? Visit the members.

Ask them what they are doing. See if they can explain the big picture. If they cannot, find a different adviser.

Often a prominent scientist will lead a big group with, say, 15 or 20 experimental systems, enabling an equal number of graduate students to study trends. These students are guaranteed to finish their degrees in a reasonable period of time. In total contrast to my own graduate student experience, they are assigned very specific problems. They take their data, report their results, and get their degrees. It all seems so easy. Should you be part of this kind of group? Again, the issue is whether the students have an inkling of the big picture. Is it only the adviser who knows what trend is being studied, while student A. is looking at rhodium, B. has a sample of ruthenium, and C. has some palladium? If the students cannot tell a good story, move on.

Choosing a Postdoctoral Position

How should you be rational about the choice of a postdoctoral position? It is essential to understand what your interests are and how they differ from the employer's. To begin, you should realize that what you actually achieved in your thesis is not especially important to your postdoctoral adviser. If you are one of the few whose thesis represents a major breakthrough, you will probably be much in demand, and will likely have few problems finding a permanent job. You probably won't ever have a postdoctoral position. Your problem may be that you will spend the next several years trying to show that your initial triumph was not a fluke. This kind of thinking has paralyzed more than a few young "geniuses" but is not an important consideration for the majority, for whom this chapter is written.

If your thesis, as is more likely, has not attracted much interest, despite your worries, you will probably find a postdoctoral slot. Employers generally feel that a postdoctoral employee is not a big risk. Unlike a graduate student, who has to be shown the ropes and whose education may absorb so much time that his or her net contribution to the progress of a project may be slight, or negative, a postdoc is a trained researcher who can be expected to be reasonably competent and not terribly demanding of supervision.

For the typical employer, a postdoc is cheap labor. At the laboratory where I work, and this is common, a postdoctoral employee receives minimal benefits. The lab pays for medical insurance but makes no contributions to a pension plan. Paid vacation is only two weeks per year, and a postdoc salary is not loaded with substantial overhead or indirect costs.

A postdoc will also be gone in two to four years. A helpful and productive one will be a blessing, no doubt, and a postdoctoral sojourn leading to a successful career can be counted a noteworthy success. But a failure by those standards is only assessed as unfortunate—not unusual, and not disastrous. Acquiring a postdoc, in short, is much like buying a piece of laboratory equipment. One assumes it will work for a while, helping to produce results. Then it will be replaced with a newer model. From the postdoctoral employer's viewpoint, signs of a candidate's viability are, accordingly: 1. an excellent thesis-research presentation—this implies that the candidate will be a good spokesperson for the supervisor's research program; 2. not having taken overly long to finish the PhD—supporting the hope that after a sojourn lasting no more than a few years, the postdoc will have produced several publications; and 3. seriousness, knowledge, engagement, and interactivity—indications that the new hire will make for a livelier, more productive, and collaborative research group.

If a postdoc candidate wants to change fields, that is not a problem but a common practice. If the candidate's thesis work did not produce a major piece of new knowledge, that is not a problem either because a postdoc is hired fundamentally to further the supervisor's research program. If a postdoc breaks new ground or does something important during his postdoctoral period, he may be offered a permanent job. If not, he will go away, and not much will have been lost. This is the employer's perspective. What should yours be?

You have three important tasks in your postdoctoral years: You must decide in what area of science to make your name. You must *finish* at least one significant project. And, you must establish your identity in the research community sufficiently to land an assistant professorship or a junior position in an industrial or government laboratory. You have little time to waste because it will not be long after you begin your postdoctoral work that you will be back on the job market.

These considerations imply that: 1. you do not want a position where your field of research is undefined. You want to get to work on a significant research project on arrival or shortly thereafter; 2. you do not want a position in which a complex technique is being perfected (which means that your chance of producing results in time for your job hunt is minimal). You want to be involved in one or several short-term projects.

If you are changing fields, you want to start your reading and learning *before* you arrive at your postdoc site. The clock starts ticking when you get to your new location. Whatever you do before you leave the nest of graduate school doesn't count, for all practical purposes. Generally, it would be wise to find a mature scientist for a postdoctoral supervisor rather than a relative novice. The reasons are the same as for a thesis professor. You do not want to be in competition for resources or credit for results. If there is only one experimental apparatus in the laboratory, or if the group computer budget is relatively thin, do you think you will be allowed to use whichever resource as much as you need? Will an adviser who has less than six years before tenure review be capable of recognizing the importance of your achieving recognition after only a year or so? There is more than a little chance not, logic dictates. Thus, unless you can find an assistant professor or junior industrial researcher who is a superstar, or at the very least, unless you can satisfy yourself that the young scientist you want to work with understands and agrees to accommodate your needs, you would probably be better off working with someone established.

Keys to success as a postdoc: Once you do take a postdoctoral position, the keys to success are: 1. *finish something;* and 2. make yourself known and useful. Your first priority as a postdoc is to have something to talk about when you go job hunting. No employer wants to hire a person who starts but cannot finish projects. Even if you have put a year and a half into developing a *very* promising method, you will lose out in the job market to your competitor whose methods may be less adventurous but who has produced a kernel of new knowledge, who has written it up and published it.

I do not recommend that you be careless in your research endeavors. Nevertheless, you should be aware that it is possible and may be desirable to publish an exciting result before the last *i*'s are dotted and *t*'s are crossed. It is possible, and relatively risk-free, if you are honest in your manuscript about the work that remains to be done. It may be desirable because someone who has a provocative story to tell, even if it is only supported by admittedly plausible evidence, will win out in the job market over someone whose very thorough effort is not far enough along to allow conclusions to be drawn. Although attention to detail is important, and publishing results that later turn out to be incorrect is anything but desirable, *finishing* projects and having a story to tell are essential. As a postdoc, under time pressure, you may have to sacrifice your desire for perfection, you may have to live with the fear that you haven't got everything just right, in order to develop a story that you can use to sell yourself. This is not cynicism but realism, and worth remembering for your entire career. The famous physicist Wolfgang Pauli is remembered for complaining ironically that the work of a young colleague "isn't even wrong." Think about *that!*

Do not be a slave to your postdoctoral adviser: If you just sit in your office working, while you are a postdoc, your supervisor will know you, but no one else will. You will get one good recommendation letter, assuming you have performed well, and that is all. If you chose a thesis adviser with good connections, he may still be able to help you find a real job after your postdoc. But what you accomplished as a graduate student does not count for much in later life, unless it is very exceptional. If your thesis adviser helps you find a job via his connections, it may be looked on as being *despite* your performance as a postdoc, and the burden on you to prove yourself in a junior, continuing position may be greater than otherwise.

What you really want to achieve as a postdoctoral researcher is to gain the respect of three or four staff members where you work who will write you good recommendations. If you are a theorist, plan on spending two or three hours weekly talking to experimentalists, and vice versa. Barge into people's labs, politely, and find out what kind of work is going on. Discover whether there are other research programs to which you can contribute. Get copies of your lab's preprints. Read them, and if you have criticisms, questions, or contributions, make them known. Every lab is eager to employ and to recommend interactive people.

If you are congenitally shy, you have a real problem, one that it would be helpful to overcome. Try to focus on the idea that positive feedback from the people you help will help you psychologically, and the recognition that their positive comments to others will advance your career.

Above all, during your postdoc years, work hard. You have only a short time to prove yourself. Do not slack off now. There is no time to waste. Your postdoctoral years represent the most intensely important period in determining whether you will have a career.

CHAPTER 4

Giving Talks

Tourist to New York passerby: "How do you get to Carnegie Hall?" Passerby to tourist: "Practice, practice, practice!"

On a job interview trip, your task is to persuade a significant fraction of the professionals who see you that they would be excited to have you as a colleague. The seminar you present is your best opportunity to convey the message that you are the person to hire. The same applies when you report on your progress after a year or two in a new position. The colleagues who know you best may already think very highly of you. But they have only a few votes. By giving a good seminar, you can add to the base of support you will need to be kept on or promoted.

Remember that few professional scientists have much time for reading. The way they learn of new and interesting work is by going to meetings and listening to seminars. If you present your work well in these venues, you will be much better able to attract a following. Having a following is an excellent form of job security.

Because oral presentations will play a vital role in your career advancement, you must take their preparation very seriously. Learning from scientists who present their talks well is a good idea. In this chapter, I hope to impart some of the basic concepts.

The Scientist as Showman

Although a seminar is not a theater piece, there are common elements. As the speaker, you are putting on a one-person show. Your listeners are investing an hour of their valuable time. Of course they want to learn something from you, but like theater goers, they expect to hear a good story, with a beginning, a middle, and an end. They don't want to squirm when you explain something poorly or wrongly, when you show a slide containing an egregious misspelling, or when the end of the hour is approaching and you obviously have a lot left to tell. Disappoint your listeners at your peril.

They might not throw tomatoes or rotten eggs, but they might dismiss you, might be unwilling to find out how good a researcher you really are—just because you put on a bad show.

The Introduction

A fundamental principle in preparing a talk is *never overestimate your audience.* No matter how gray their beards, no matter how many papers a few might have published in your field, those frightening-looking people in the audience *want* a complete performance. They want you to say what is important in the area of interest, particularly if what is important happens to be their own work! They don't mind hearing things they already understand—it makes people feel good to understand something.

The opening lines of a talk set the tone, make a first impression. The main impressions you want to make are that: 1. you know your field; 2. you are possessed of the scientific curiosity that will make you a valuable colleague; 3. you enjoy doing research; and 4. you plan to convey some useful and interesting information. Tell the audience what the theme of your presentation is; or tell them that your work was undertaken to resolve a particular controversy, and why it is an important one; or tell them that you have demonstrated a novel technique, which permits access to new and useful information.

Do not simply launch into a discussion of the experiment or calculations that you did. Establish the context of your research to the degree that time will permit; give an overview of the novel technique, ideas, or shortcuts you have employed; and possibly, intimate what the most important conclusions are. ("These measurements, as you will see, confirm the long-standing, but until now unproven, predictions in Feibelman's early, brilliant paper.")

This done, you can go on to discuss the specifics. If you are giving an hour's talk, you will want to expand on your introductory remarks before launching into the details of your own work. In a ten-minute paper at a large meeting, a one- or two-slide introduction may be enough.

Stagecraft

Be aware of the importance of your demeanor, particularly your air of self-confidence. If you speak almost inaudibly, it will be assumed that you lack confidence in, or do not understand, what you are saying. If your presentation is too low-key, you may convey the idea that you are not enthusiastic about your work, or perhaps about research in general. Scientists are like terriers, trained to chase down and pick apart reasoning that is not rigorous. If you appear confident, your presentation is more likely to be accepted at face value. If not, you can expect to be fielding insistent questions early on and may never get to finish your talk. Alternatively, you may see people walking out of the seminar room. If you are interviewing for a job, that could be rather disconcerting.

Time is of the essence when you are giving a talk. You must plan your presentations and rehearse them, to ensure that you will be able to finish before your time is up, or at least to be sure you will have conveyed the main ideas by the time the bell rings. You can easily determine in practice sessions how long it takes you to present an average slide. This will make it easy to fix an upper limit on the number of slides to prepare for a given time slot. Personally, I can discuss six or at most seven slides in ten minutes. If I prepare more than that, I know that my talk will be breathless and that my audience will absorb little. They may well respond to a talk too crammed with information as a "snow job," an attempt to disguise the flaws in your work by overwhelming your listeners with words and figures. Designing a modular talk is a good idea. After your introductory module, you present several complete information packages in sequence. That way, if you see your time running low, you can excuse yourself for leaving out the last module and skip ahead to your summary.

Don't Try Their Patience

One of the first lessons students learn about giving a talk is to "prepare an outline." Many of them are also apparently taught to begin with a slide that gives "an outline of my talk." I often find these slides a waste of time, if not downright silly, and would like to dwell here on the structure of a talk, not just to help you, but hoping that I will have to sit through fewer outline slides in the future.

Have you read a novel recently, or seen a play that started with an outline of the plot? When a political candidate gives a speech, does he put his outline on a chart? Of course not, and in general, neither should you. You certainly should outline your presentation in the privacy of your office. But in giving your talk, you should just tell a story. Its structure should be organic, invisible. Your listeners should be propelled from idea to idea with the same sense of inevitability they feel on hearing a Bach fugue.

At meetings of the American Physical Society (large meetings), contributed papers are allotted ten minutes plus two for questions and discussion. Thus, I can present six or at most seven slides in such a talk. What message do I convey if Slide 1 is "The title of my talk," and "these are the names of my collaborators, and I want to thank the Department of Energy for my funding," and then Slide 2 continues with "I will begin my talk with a brief introduction. Then I'll discuss our experimental apparatus. Following that, I'll present my results for system X, and finally, I'll end with some conclusions." All right, this is something of an exaggeration, but it is not an enormous one. What it conveys is that "I don't have much to say, so I'll throw away most of my time telling you how I planned my talk and who my friends are, leaving little time for any discussion of what I have learned." If you have nothing to say, you would be better off not giving a talk. If you do opt to speak, you do yourself an injustice not using virtually all your time to present your ideas and results.

One of the wonderful abilities people have is to take in different information with their eyes and ears, simultaneously. If you have collaborators not announced as coauthors and a funding agency, do acknowledge them on your title slide (Fig. 1), but do not waste time reading their names. Someday, when you are a professor and are trying to place your students, then you can mention their names and good qualities (usually at the end of your seminar). Now, however, *you* are the person you are trying to sell. Acknowledging your coworkers is important but should not be overdone.

What you want to convey in your introduction, while your title slide is on the screen, is what got you interested in the material you are about to present, or perhaps why researchers in your field are interested, or why the community as a whole should pay attention. What you actually say should be geared not just to the subject of your work but also to the nature of your audience. Clearly, if you are giving a ten-minute presentation to experts in your field, you should dispense with remarks of too general and introductory a nature. On the other hand, if you are giving a colloquium to an audience including professionals expert in areas other than yours and students, then a long introduction is essential.

Stimulative Properties of Elixir X
I. M. Balding

SUPERVISOR:
Prof. A. Barber

ADDITIONAL HELP FROM FELLOW POSTDOCS:
Sam Son (dendrite growth)
D. Lila (cutting tools)

FUNDING:
Nat'l Hair Council

FIGURE 1

Technical Matters

Attention to the technical aspects of talk preparation can make the difference between a good seminar and an excellent one. Experimental solid-state physicists always seem to show a slide featuring a schematic or, God help us, a photograph of their apparatus. Occasionally, there is good reason for such a slide. More often than not, it is a waste of time. "Get to the ideas!" I think in these cases. In putting together the body of your talk, try to recognize digressions for what they are. If there is a good reason for showing an equipment slide, if it explains a novel technique, then do it. If the measurement method is standard, if the slide only proves that your lab isn't empty, that you didn't make up your "results," forget it. Nobody minds a short, informative talk. Don't pad your presentation by design or by inattention to preparation.

Theoretical physicists, particularly inexperienced ones, often show slides covered with equations. (Molecular biologists show DNA sequences.) Except in very special cases, such as meetings of specialists devoted to technical advances, this is a bad idea. The audience cannot assimilate more than a small amount of information in an hour, to say nothing of ten minutes. A talk comprising detailed, technical slides is likely to be received as a deliberate attempt to persuade the listeners that because the material being presented is so complex as to be incomprehensible, it should be looked on as important. Save this for after your Nobel prize. Then, most of your audience will be afraid to reveal that they have no clue as to what you have done, or that they despise your snow job. For now, you need to please your audience, not beat them into submission. Put yourself in the place of an experimentalist among your listeners. Why would he want to hire you? There is an outside chance he would act in your favor because a colleague who actually understood your equations told him that they are important. More likely, he would prefer someone he thought he could talk to. To communicate with him, you need to convey not the details of your math but the basic concepts, the approximations, the results, and the predictions. Think about that. Then throw away that slide cluttered with superscripts and subscripts.

Slides: A few ideas on laptop presentations are certainly in order. When I see a beautifully prepared, multicolored slide, what first goes through my mind is, "this guy obviously doesn't have enough to do." Granted, modern technology makes the preparation of professional-looking presentations relatively easy. Nevertheless, you do not want to give the impression that thinking about how your slides look is more important to you than what they say. If you are preparing a talk for a group of laymen—e.g., upper management or an army general—by all means make your visual material spiffy. But for your fellow scientists, go easy on the "professional" look. Remember that many of them have been driving a beloved old car for years, and the same reverse snobbism that keeps them in their clunkers probably also affects their impression of your slides.

This, I hasten to add, does not mean that your slides should be prepared thoughtlessly. For the most part, they should contain a figure or two, a "cartoon," and simple text. Showing slide after slide of bullet points risks inducing yawns, and is not recommended.

Go easy on animation. It is disconcerting to see words fly onto the screen from every which direction. Animation also incurs a risk: Should you have to back up to mention something you forgot to say, your words will be flying out again and then, possibly to giggles, back in. In a similar vein, do not overdo the use of fonts and colored text, which tend to tax the viewer's eye. Do try to leave a reasonable amount of white space on each slide; that tends to be relaxing.

Use large fonts! This has two advantages. One is that people in the back of the room, close enough to the door that they can escape inconspicuously, can read what you've written and might be persuaded to stay. The other is that it limits the amount of material you can fit on a page. You don't want a lot.

You might be wondering how large is large enough. To decide, take your laptop and a projector to a seminar room. Look at your slides from the back. Can you read them? While you are there, notice whether your color scheme provides sufficient contrast. Can you read your light blue letters against a white background? Black might be better.

Summary

By now, I hope you have realized that this chapter is organized as a seminar on seminars, and I would like to reiterate the main ideas:

1. Your seminar is a performance. It needs to be carefully planned and thoroughly rehearsed.
2. Present yourself confidently. Act as though you have enjoyed doing your research and that your results are exciting to you.
3. Respect your audience. They are spending an hour to hear you. They want to understand what you have to say, even if your specialty is not theirs. They do not want to be "snowed," nor do they want to be treated as experts in a field where they really are not.

4. Do not waste your time with filler. Make sure each slide pushes your story forward. If your talk is a bit too short, no one will object.

5. Make your visual aids pleasing to the eye without too much of a Madison Avenue look.

Thanks for your attention!

ADDITIONAL READING

Garland, J. C. "Advice to Beginning Physics Speakers." *Physics Today* 44, 42 (1991).

Booth, Vernon. *Writing a Scientific Paper and Speaking at Scientific Meetings.* 2nd ed. New York: Cambridge University Press, 1993.

Alley, Michael. *The Craft of Scientific Presentations: Critical Steps to Succeed and Critical Errors to Avoid.* New York: Springer-Verlag, 2003.

CHAPTER 5

Writing Papers
Publishing Without Perishing

The negative connotation of the cliché *publish or perish* is seriously misplaced. Publication is a key component of your research efforts. It is widely accepted that a scientific endeavor is not complete until it has been written up. The exercise of putting your reasoning down on paper will frequently lead you to refine your thoughts, to detect flaws in your arguments, and perhaps to realize that your work has wider significance than you had originally imagined. Publication also has strategic significance. As a beginning scientist, not only do you work long hours for low pay, but your job security is anything but assured. To succeed, you must make your talents well known and widely appreciated.

Publishing provides you with an important way to accomplish that. Your papers, on public view around the world, represent not only your product but also your résumé. Compelling, thoughtful, well-written articles are timeless advertisements for yourself. You can imagine that a sloppy résumé is not worth preparing. A premature or slapdash publication is far worse. It will remain available to readers indefinitely. These thoughts raise the two basic questions addressed in the present chapter: *When* should one write a paper, and *how* should one write it?

Timing

Generally, articles are written too soon in response to the fear that one's competitors will publish first or as a result of intellectual laziness (i.e., inattention to important details). Papers are written too late because of the fear of publishing a blunder or because of writer's block. Overcoming these fears and frailties is necessary for *everyone* in science. At the very least, the knowledge that they are not yours alone may help you deal with them. (Read Carl Djerassi's novel *Cantor's Dilemma* [New York: Penguin Books, 1991] for a poignant exposition of the problem of when and what to publish.)

Planning your research as a series of relatively short, complete projects (cf. Chapter 9) is the best way to achieve a disciplined publication schedule, one that serves your interests in scientific priority, self-advertisement, and job security. Even though you are working toward an important long-term goal, you report each project as an independent piece of work that has produced a new kernel of knowledge (only half-jokingly a "publon," a quantum of publication*). In the introduction to each paper of a series, you place the work reported in the context of the long-term goal, to which you thereby lay claim, and you explain how the present results take you a step closer. If your project turns out to be as significant as you had hoped, after you have published several papers in the series, no doubt you will be asked to write a review. *This* will provide you with an appropriate forum for a long, definitive article, one that will be widely referred to and will help to make your name in science.

There are many advantages to writing up your work as a series of short papers. Managers and funding agencies need concrete evidence that they have hired personnel and spent money wisely. Nothing is more helpful in this regard than the list of publications their wisdom has fostered. Of course, they will be pleased if you eventually realize a long-term research goal. However,

---

* The concept of the "publon" emerged from the graduate student minds of M. J. Weber, now at the University of Virginia, and W. Eckhart, now at the Salk Institute.

funding cycles are typically two or three years (cf. Chapter 8), and renewal of junior scientific positions occurs on a similar time scale. Therefore, deans, research directors, and contract managers cannot wait for your long-term dreams to come true. They need published evidence of your progress on an ongoing basis.

By writing numerous, relatively short articles, you can keep your name in the spotlight. The titles, abstracts, and authorship of your new papers will show up in electronic databases, generally updated weekly. Such search engines as scholar.google.com, www.osti.gov/eprints, and www.scirus.com will readily lead the community to manuscripts you have posted on arXiv.org, precedings .nature.com, or any of a host of other preprint servers. The number of citations of a long publication list increases more rapidly than that of a short list.

You mustn't be overly cynical about these facts of scientific life. If you attempt to achieve name-recognition by padding your publication list with repetitive papers, your efforts will soon reap scorn rather than admiration. Still, the little admiration you gain for publishing an awesome magnum opus in a single paper is surely not worth the risk that this publication strategy poses to your job security.

If you publish frequently, you are less likely to be "scooped." The longer you hold back reporting your results, particularly if they are important, the greater the chance some other group will beat you into print. You do need to develop an appreciation for when a piece of work is complete enough to be written up. If the logic of a manuscript is clearly missing an important piece of confirmatory evidence, submitting it to a journal is likely to cause you endless, painful interactions with referees. This is the time to hold back. (Among other problems, the referees may very well be your competitors. Their own publication strategy is likely to be affected by their appreciation of where your incomplete work stands.) On the other hand, if you *have* completed a project, the sooner you get it into the hands of a journal, the better the chances are that you will get credit for your accomplishment.

Writing a paper that presents one new idea or result is much easier than writing a long, complex article. This is a reasonable way to address the problem of writer's block. Much of the introduction to a shorter paper can be prepared, at least mentally, when the long-term research project is originally proposed. The organization of a paper is simpler if there is not so much material to present, and it is also relatively easy to explain the conclusions in that case.

Referees are generally busy people and prefer to review short papers. You are likely to receive a more thoughtful and positive report on a short manuscript than on a long one. Shorter papers are of course not

only easier on referees. They also can be read and as-
similated more easily by the scientific community at
large.

Writing up individual kernels of new research
should have some appeal for the perfectionist. It is easier to get everything right when one is dealing with a
small project than when publishing the results of a
major, complex effort.

Eventually, of course, all the significant details of a
research project need to be reported in an archival
journal so that others may repeat and confirm the va-
lidity of the new science. Writing such technical papers
is an important exercise, and one that will win you
credit from your peers if you do it well. On the other
hand, in most cases the writing of such papers can be
carried out at leisure.

Writing Compelling Papers

A journal article should present a careful and relatively
complete account of your research. However, it is all
too easy to write an accurate description of your work
that attracts no attention and that adds little to your
scientific reputation, *even when your results are significant.* Learning to write articles that people will
read and remember will make you a more effective scientist. It will also enhance your chances for survival as
a researcher.

The structure of a news article is a good model to follow in preparing a scientific publication. Newspaper readers, like your research colleagues, rarely have much time for acquiring new information. This is just the reason that news articles present a story several times, in increasing levels of detail. Their headlines, equivalent to the titles of your scientific papers, are there to draw readers in by providing a succinct description of what is noteworthy. Scientists attempting to keep up in a world of information overload often do no more than skim the tables of contents of the leading journals in their field or conduct electronic keyword searches. You can help direct them to your new paper by taking the time to prepare an accurate and compelling title, concise yet incorporating the most important keywords. ("Cute" should be avoided, as a rule.)

The abstract of a paper corresponds to the first paragraph of a news item. It summarizes the main information, what the important results are, and what methods you used to obtain them. Numerous journals place a word limit (e.g., 75 words) on the abstract. It is a good idea to impose such a limit on yourself whether or not the journal does. An abstract that is brief and to the point has a better chance of being read. A wordy one, which reads like the introduction to or the body of a paper, will lose readers.

As in the case of titles, it is worth remembering that abstracts circulate more widely than the papers they summarize. They are the first item to pop up when one searches journal content and are generally available without charge, even when seeing a full article requires a subscription. A well-written abstract may thus make the difference between someone's downloading your full text or emailing you for a copy, rather than just moving on.

The introduction to a paper is where you tell your story, possibly illustrating the text with an important figure or some key results, but without going into great detail. Here is where you want to explain why your project was an important one to undertake and how your results make a difference to the way we understand the world. Many busy scientists read only the introduction and conclusion sections of papers, leaving the technical details for another time. Therefore, it is a good idea to highlight your results—for example, by placing your most important figure in the introduction. Even if your readers never take the time to plow through the complete description of your work in the body of your paper, they may think enough of the information in your introduction to make sure to catch your talk at the next scientific meeting.

Virtually everyone finds that writing the introduction to a paper is the most difficult task. It is easy to report the procedures you followed and to describe the data you obtained. The hard part of paper writing is drawing the reader in. My solution to this problem is to start thinking about the first paragraph of an article *when I begin a project rather than when I complete it*. I would not embark on a scientific effort if I didn't think it was important and that my work would answer a question of rather wide interest. The reasons that I found the project in question interesting enough to work on provide half the material I need for my introduction. The remainder is a summary of my key results. The decision to start writing a paper is generally based on recognizing that a kernel of knowledge has been produced. In my introduction, I want to let my reader know what this new information is, in a nutshell, and why it is worth reading about. Sitting at the word processor, I imagine I am on the phone with a scientist friend whom I haven't spoken to in some time. He asks me what I have been doing recently. I write down my imagined response. If, when you try this, you feel an attack of writer's block coming on, turn on a recording device and actually call a friend. It works.

Incidentally, if you know why you have carried out a scientific project and what makes your results interesting, there is no reason that your paper should start with an inane cliché, such as, "Recently there has been a resurgence of interest in . . . (whatever the topic)," which bothers me every time I see it. If you have been working on a project for several months or a year solely because *other* people are interested in it, you have a lot to learn about problem selection. (In this case, see Chapter 9 for some help. Do not pass go. Do not collect your next paycheck.) Before you start on a research effort, you must understand why it is important, and in the introduction to your publication on the subject, this is just what you need to explain.

In writing your introduction, as well as the body of your paper, it is essential to place your work in context, not only by explaining what you did and why but also by citing the relevant literature. This is important, not only to provide your readers with a way of understanding your area of research, but also because your scientific colleagues are very eager to get credit for their achievements. (This is not just vanity. Scientists' careers are built on the perceived importance or usefulness of their research results.) You have much to gain and little to lose by scrupulously citing your competitors' work. I said above that many busy scientists read only the introduction and conclusion sections of papers. Even more move directly from the title and abstract to the references, to see if their work is cited. If someone's papers are not mentioned there but should be, you risk losing a potential friend or at least some respect.

I would add that an excellent way to keep up with developments in your field is to check, from time to time, who is citing your own papers. A "citations index," such as is available on the ISI Web of Science[SM], makes this an easy task. Bear in mind as you do this that if checking citations is how people in your field keep up, an article you have written that fails to cite their work is more likely to go unnoticed.

In revising and editing your article before submitting it, you should constantly be asking yourself if you have dealt with all the loose ends in your logic. Are there arguments you have thought about and used but not written into your text? Are you wishy-washy about inferences you have drawn, instead of forceful, because there are missing links in the logic? If so, you either need to work a little longer before writing your paper, or be forthright about what is conjecture and what has actually been proven. Even if the referee does not catch the weak points of your manuscript, you must not forget that your paper will be on public view for a long time. Intellectual honesty is accordingly a very good policy. This is not to say you should be such a perfectionist that you never feel comfortable declaring a project done and ready to be published, but rather that you should own up, in print, to what you think might be weak links in your reasoning. This is a service to the community, in that it points to further research directions. It shows the world that you are a thoughtful and forthright individual. Importantly, it also provides you an out if your reasoning is later shown to be incorrect.

The format of the body of a paper is often dictated by the journal where it will be submitted. Within the journal's constraints, however, the key to organizing your work is to make your text read like a story. Often it is a good idea to relegate detailed discussion of a technical aspect of the work to an appendix. That way, experts or interested parties can try to understand your arguments in full detail, whereas others do not have to guess how much of the text to skip to move on to the next idea.

Keep in mind that the function of a journal article is to communicate, not simply to indicate how wonderful your results are. In principle, a paper should provide enough information that an interested reader would be able to reproduce your work. It is your responsibility to ensure that the necessary information is made available, at the same time as you try to make your paper as snappy and readable as you can.

Snappy Papers

In archaic times, say 30 years ago, you generally had to write your papers as though the work had actually been done by someone else. You were discouraged from using the personal pronoun "I" in favor of "we" or, even worse, "one." Journals seemed to require writing papers in the passive mood, as in "the data were obtained using the following novel method" rather than "I developed the following novel method to obtain the data." More recently, it has become possible to drop the phoniness of this style and to reveal in your writing that *you* actually did the work you are reporting. I greatly prefer the more straightforward style and recommend that you use it.

People of a mathematical bent often connect the sentences in their papers with such words as *now, then, thus, however, therefore, whence, hence,* and so forth. If you want your text to be readable to the non-pedantic, you should be very sparing in using them. Go over your first draft and challenge yourself to see how many of these connectives you can remove without undermining the logic of your argument.

In this era of speedy desktop computers and full-featured graphics programs, there are few excuses for omitting evocative figures from a paper. A picture may be worth more than a thousand words in a scientific article, particularly if the thousand words are not read, but the thoughtfully prepared figure is examined and the information it reports absorbed. This does mean it is important not to prepare figures that are too cluttered. If they offend the eye, they may be ignored along with the thousands of words.

Some journals restrict the length of articles. This typically forces one to go back through the first draft of a manuscript to rewrite more economically. In preparing the first draft, it is a good idea to be as generous as possible with words. You should write down everything that comes to mind as relevant. This may not be easy but helps get all the logic on paper. (Again, get out the voice recorder if you tend to be stingy with words.) If you have written a copious text, the exercise of cutting back may be more difficult but is less likely to lead to a paper whose flow is compromised by the absence of something important. I recommend the approach of writing generously and then editing severely in all cases —that is, whether or not the journal in question imposes restrictions on manuscript length. The exercise of rewriting as concisely as possible leads to more readable text and thus to text that is read more widely.

As in the preparation of a seminar, the last section of a paper should provide not just a summary of the results reported but also some idea of how they might affect the direction of future research. The goal of the conclusions section is to leave your reader thinking about how your work affects his or her own research plans. Good science opens new doors.

Referees

Last, because arguments with journal referees can take many months to settle, and can be very frustrating, it is a good idea to forestall them by having your manuscripts reviewed locally, by one or two of your colleagues, before submission. If you have chosen your local reviewer well, you may discover the weak points in your article in a matter of days rather than months. If English is not your mother tongue (and if you are writing for an English-language journal) it is even more important to have your paper reviewed and edited by a colleague, one whose English is near perfect. Your readers, including your journal's referees, are human and thus impatient to some degree. The easier you can make their task, the better will be their response to your efforts.

Incidentally, as one who referees many papers, I much prefer receiving a cogent, well-written manuscript that I can learn from than the other kind. A paper that I enjoy reading disposes me favorably toward the author. Your referee may be your paper's most careful reader ever. Making a good impression on this anonymous potential employer is not a bad idea!

If your referee does have serious complaints about your article, getting angry is not a productive response. A better idea is to consider why this thoughtful expert did not follow your argument and agree with it. If on reflection you believe that your results are correct and that the referee has simply misunderstood them, it is likely that spending some time revising your text will not only persuade the referee to recommend that your paper be published but will also ultimately make your ideas less confusing to your journal's general readership.

ADDITIONAL READING

Carter, Sylvester P. *Writing for Your Peers: The Primary Journal Paper.* New York: Praeger, 1987.

Alley, Michael. *The Craft of Scientific Writing.* 3rd ed. New York: Springer Science and Business Media, 1996.

Booth, Vernon. *Communicating in Science: Writing a Scientific Paper and Speaking at Scientific Meetings.* 2nd ed. New York: Cambridge University Press, 1993.

CHAPTER 6

From Here to Tenure
Choosing a Career Path

As a scientist, your goals are to make exciting discoveries, to change the way your colleagues and maybe even the public at large view the world, and generally to improve people's lives. However, need I remind you, you will remain a human being, with human needs, even while you are pushing back the frontiers of ignorance. No matter how romantically you view your role in research, you will not be happy without a secure, well-paid job. You will want help in accomplishing your research goals and recognition for your achievements. You will probably want to see your family on a regular basis and, more generally, to have enough free time to engage in activities outside your professional life.

It is all too easy to lock yourself into a situation where one or more of such basic desires will not be satisfied. This may adversely affect your productivity, your family life, and your ability to enjoy yourself. Thus it is important to consider rationally, and in advance, not only the benefits and disadvantages of the various kinds of scientific positions—academic, industrial, and governmental—but also the merits of the different roads to permanent employment.

Economic conditions may limit your choices, but if you are fortunate enough to have more than one job possibility, this exercise will save you considerable stress. It may have a significant effect on your financial well-being. It may save your marriage. I harbor a secret hope: If enough of you start to act rationally, the system may eventually be rationalized.

It is only natural to adopt as role models the people one encounters in one's formative years. For this reason, many—perhaps most of us—finish graduate school dreaming of an academic career. For some, the academic life may be ideal. For many, it is not. Even if being a professor *is* the right goal, however, it is far from clear that rising up the academic ladder is the most desirable way to get there. My recommendations and the reasons for them are the subject of what follows.

The Pluses and Minuses of a Job in Academia

The idea that a university is an ivory tower is commonplace. The academic freedom embodied in the granting of tenure was originally supposed to protect the professoriat from political repercussions against expressions of minority views of the world. However, tenure is in itself a uniquely desirable and economically significant benefit.

Who wouldn't want the ultimate in job security? As a tenured professor, if you fulfill minimal performance requirements (e.g., teaching a class every semester) and maintain at least minimal moral standards (love affairs with your students are sometimes frowned upon), and if your university doesn't shut down your department entirely in response to severe economic stress, you have a guaranteed paycheck. In fact, universities have long since recognized the economic significance of tenure. University salaries would certainly have to be higher if professors were subject to being laid off.

Tenure is a form of financial independence and thus conveys corollary benefits. A university professor chooses research topics and collaborators at will. No boss is empowered to say what to work on or to decide who will work with whom. In principle, the pace of research is also up to the professor. If energetic and ambitious, an established professor, together with a group of students and postdocs, may produce a dozen publications a year, or more. A "scholar" may publish many fewer, might be poorly funded, and may not have much of a group. The department chair or the dean may complain, but the scholarly professor will still receive a paycheck.

Although tenure and its corollaries are the unique benefits of a professorship, they are far from the only attractive features of the job. Professors can anticipate the respect not only of class after class of students, who pay a great deal of money to be exposed to what they have to say, but also of the community at large.

Typically, professors are free to sell their services as consultants, perhaps one day per week, to supplement their salary. Many science professors found private companies to develop the fruits of their research and sell them for their own profit. Others write textbooks on university time and pay, and then are allowed to reap the royalties for themselves.

Because classes are held only nine months of the year, the remaining three are in principle a very long annual vacation or at worst, unprogrammed time. Sabbaticals are typically part of a university contract. Every several years, professors can look forward to six months or a year at a distant and often exciting location where they can recharge their intellectual batteries, learn a new field, write a book, or basically do what they please—and get paid for it!

Given that the job has all these wonderful benefits, you might be surprised that many professors complain about the demands of their work and that many scientists are happy not to be members of the professoriat. What, then, are the disadvantages of living in the ivory tower?

Probably the most widespread complaint is that a professor rarely has time to set foot in the lab and to do the scientific research that used to be so much fun. Professors have so many responsibilities and have to work so hard to fulfill them that their scientific work is mostly vicarious—it's the students and postdocs who do the hands-on research. To say the least, professors end up with little time for themselves. There are thankfully few tenured individuals who cynically view their permanent slot as an opportunity to do nothing (although there is generally more than enough "dead wood" in a department to embitter the assistant professor not promoted to tenure). The professors I know work many more than eight hours a day and rarely take more than a week or two of vacation each year, even though in principle they could take much more.

A professorship is effectively several jobs rolled into one. A professor is of course a teacher. Although there are many stories of professors whose lecture notes are yellowed with age, taking the job of teaching seriously means devoting considerable effort to making classes coherent, informative, and up-to-date. One needs to prepare homework sets and exams and to develop meaningful lab exercises. One must also spend time with students during office hours. A professor is expected to be a good departmental citizen. This means attending a significant number of meetings to decide policies and to discuss hiring and promotions. The ambitious professor spends a great deal of time as a manager. This means writing grant proposals, traveling to Washington to meet with grant administrators, fighting for lab space, hiring and firing students and postdocs, and so forth. Being an active scientific citizen, which includes refereeing manuscripts and grant proposals, preparing and giving lectures at other institutions, and attending conferences, also absorbs hours. Consulting and textbook writing come on top of that. It does not take a genius to see that professors have little time for reading a novel or playing with the kids.

A job with many demands provides many opportunities for frustration. When economic times are tough, the chances of getting a proposal funded or renewed are reduced. If you have no grant money, you cannot afford to pay students and postdocs. If you cannot spare much time to do research yourself, this means your research program will grind to a halt. Your ensuing lack of productivity will then make it harder for you to acquire funding in the future, a most unpleasant feedback mechanism. Apart from keeping yourself alive as a researcher, if your funding dries up, you may find yourself struggling to make ends meet. Typically, a university salary is only paid during the academic year, and if you are not bringing in substantial outside money, your nine months' pay will not be particularly generous. (The university reasons that you are unlikely to give up your sinecure for less than a major pay increase, something a poorly funded professor is unlikely to be offered elsewhere.) Your application for a research contract will therefore generally include a request for "summer salary"; most universities allow you to receive two months' pay from grants. This makes getting funded intensely important to your pocketbook. If you succeed, your annual pay can increase by better than 20 percent. If you don't, you may wonder why you are working so hard.

Interacting with students can be a great pleasure but is often very stressful. As a teacher, you will have to deal with insistent people who want to know why their exam grades were so poor and who want private help to understand the material you have been presenting.

You will have to deal with students who cheat on tests and with premeds who have no interest in anything but grades. Only some of your graduate students will really contribute to your research. Others will break your equipment, contaminate your samples, and install bugs in your computer programs. Some postdocs (particularly those who haven't read this book!) will flounder for a year or two, will be bitter about their inability to find a job, and will complain publicly about your guidance.

Your academic freedom is certainly a great benefit, but what about that of your colleagues? In some departments, the various groups talk to each other. However, this situation is far from guaranteed. Because there is effectively no management in a university, professors tend to work independently. There is no particular reward for collaboration. This is very different from a national or industrial lab, where the job description includes helping to promote the efforts of one's professional colleagues.

Assistant professorhood: If after this litany of disadvantages, you still want to be a tenured professor, there remains the question of how to attain such a position. The most direct route is to work your way up from the bottom, that is, to start as an assistant professor and to be promoted. I heartily recommend that you avoid this path if at all possible.

As an assistant professor, you suffer most of the disadvantages and have few of the benefits of a tenured academic position. Not only do you have to teach, but unlike your senior colleagues, you haven't got sheaves of lecture notes from yesteryear. You start from scratch—which means devoting many, many hours of preparation for each hour you spend in the classroom. The same is true when it comes to preparing homework assignments and exam questions.

Although being responsible about your teaching duties is necessary for you to win promotion to tenure, at a research-oriented university, it is far from sufficient. You will certainly be judged on your ability to bring in grant money. Although you will have to publish to avoid perishing, you will also have to get funded to survive. This means you will be learning the ropes of grant writing at the same time as you are trying to establish a research effort and desperately need to produce some results.

Your salary as an assistant professor, as for all professors, will not only reflect your seniority, or in this case your lack of it, but also your success at bringing in outside money. Since you are just starting out, you will have had no such success. Therefore, your salary will be miserly to poor. If you are such an exciting prospect that you have managed to land an assistant professorship at a major private university with a fancy reputation, your salary may be even worse. Such a university can expect you to accept lower pay in return for the snob appeal of its name on your résumé. It can also offer significantly reduced opportunity if any for promotion to tenure, on the perhaps correct assumption that its name is worth more to you than job security.

Unhappily, whereas full professors might accept lower pay in return for the grant of tenure, assistant professors are expected to take the low pay without the compensation of a secure position. Responding to the American Association of University Professors' (AAUP) efforts to protect you against exploitation, most schools adhere to the policy that an assistant professor who hasn't been granted tenure after seven years must be fired. Thus, ironically, thanks to a labor organization that purports to represent your interests, you will lose your job if you are not promoted!

There *are* pleasures to working as an assistant professor. Teaching and interacting with students can be exciting. The university environment is in itself very stimulating. There are certainly more kinds of people with more diverse interests than in any industrial lab. You do get respect from the community. On the other hand, the price of being an assistant professor is much too high. The hours are long, the pay is terrible, and the job security is bad. After your years of study for a PhD and further years as a postdoctoral apprentice, you will probably be about thirty years old. You'll probably be starting a family. Your former colleagues who went to engineering or business school will be making their way in the world, earning good salaries, and having time to participate in activities outside their jobs. Do you want to be working 16 hours a day for half what they are earning, on the chance that after five or six years your department may give you tenure? If enough of you answer "no," maybe the job conditions will improve. Until then, I recommend that you find a position in an industrial or government research lab. There you can establish a reputation with much less pain, as discussed below, and, reputation in hand, can start at the top in a university job, if that is still what you want.

Industrial and Government Research Positions

Research jobs in industry or at government labs have some serious disadvantages but many benefits relative to university professorships. At some of the national labs, there are tenured research positions, but for the most part tenure is not offered outside the framework of a university. You can be laid off for a variety of reasons if you work for private industry, of course, but also if you are employed at a government lab.

There is no doubt that tenure is a valuable benefit. However, you should remember that your real job security as a scientist is the recognition and approval of your peers around the world. If your published research is admired and used by fellow scientists everywhere, you have little to fear. One day you may have to change job locations, but unemployment should not be a worry. Industrial and government labs provide an environment where it is relatively easy to establish a scientific résumé. Thus, if you are competent, the issue of tenure ends up being relatively insignificant. (Incidentally, the reluctance of the managers who hired you to admit that they made a mistake provides an additional, if melancholy, form of job security at a research lab. Firing you after six or seven years if you are not promoted is not built into the system as at a university.)

The most important advantage of working in a research lab, whether industrial or governmental, is that your job description is relatively simple. You are expected to be a scientific leader, to advance knowledge in one or more areas of importance to your employer, and to make yourself useful to your fellow employees. The modern world being what it is, you can also anticipate being asked to help bring in funding. Because your main task is to produce results that will sooner or later benefit stockholders or the taxpayer, your lab will *want* to provide you with the necessary hardware (within budgetary constraints, of course), and if your work has a high priority, this hardware will be in the form of the latest and highest power models. For example, while your university colleagues are writing lengthy proposals to buy a work station, at a research lab you will be struggling to keep up with the latest upgrade to the multiteraflop, massively parallel processor. You get the idea.

Because your job description at a research lab is simple, you can perform up to expectations without working unusually long hours. As a professional, you will certainly find yourself working long days occasionally, when you are on the threshold of an exciting result, or when you have to submit an article by a certain deadline. However, you will not be spending half your time doing work that is necessary but not sufficient for your survival (i.e., teaching, explaining to students why they got a D on your last exam, etc.). You will therefore have time to help your spouse with dinner, to read a novel, to see your kids' school play, or to be a soccer coach. You won't have historians, specialists in Russian literature, or bassoon professors for colleagues, so you will have to make more effort to enhance your cultural life than at a university. On the other hand, you will have more time to spend with friends from outside the workplace.

A research lab is a *managed* environment. We'll consider the downside of living with managers momentarily.

The advantages are that management monitors the functioning of the lab and has the power to make it work better, and also that management is paid to do bureaucratic dirty work that would otherwise find its way to your in-box. At a government or industrial lab, significant portions of annual pay raises are awarded for merit rather than for having been employed one more year. There is unavoidably some arbitrariness and subjectivity in the annual performance reviews by which merit pay is determined. Nevertheless, the fact that a group seriously considers whether your work is achieving recognition and deserves a special reward, whether you and your colleagues are interactive, and whether support personnel are doing their jobs makes the atmosphere at an industrial or government lab enormously different from a university's. Employees who know that their attitudes and performance will make a difference to their paychecks take collaboration more seriously. At a research lab, you will find librarians who offer to photocopy articles for you and who will do electronic literature searches; you will find computer support personnel who want to advance their own careers by helping you make your computer programs more efficient, and who will hold your hand while you are learning a new system. You will find groups of professional scientists addressing the same complex problem from several different perspectives, groups who meet to share new results and think up succeeding experiments. At a university, such collegiality is rarer.

There are many ways that management can make your life less rather than more pleasant. Abrupt changes in corporate or congressional priorities may be imposed on you if you work at a commercial or government lab. You may have to redirect your research plans, or even terminate a project before it is completed, because of your company's poor earnings or because of political changes in Washington. Your research progress may be impeded by incessant demands to take Internet or live courses—on protection of intellectual property, "export control," shop safety, types of fire extinguishers, and . . . you name it. Heavy-handed scientific managers may insist that it is more important for you to work on their latest (hare-brained?) idea than your own. They may reinforce this by refusing to buy the equipment you want for your own purposes. They may insist that you put their name on your papers or patent applications. Or, conversely, your supervisor may have little knowledge of your field and try to compensate by requiring you to write reports on a too-frequent basis. Management may badger you with the latest buzzwords or theories

to emerge from business schools* instead of inspiring
you with rewards in the form of new instruments for
your lab and more money in your bank account. Lastly,
personality conflicts with someone who has the power
to fire you, to determine whether you can give an invited paper in a faraway place, and to control the size
of your paycheck can cause you plenty of grief.

Obviously, if you work in a managed lab, you need
to have some feeling that you will not be subject to a
too–heavy hand. A bigger lab, for example, will provide
you more freedom to correct a bad situation than a
smaller one would. At a large lab, if you just can't get
along with your supervisor, there may be several other
groups who would be happy to benefit from your wisdom and whose supervisors would be easier to deal
with. As your reputation grows, of course, your management will look to you for new ideas and be less
likely to suggest that you change directions. In a sense,
this is another aspect of the reward system in a managed environment. The more credibly you play the role

---

* "Empowerment," interpreted by many in the trenches as the
ability to be blamed rather than heard, and "thinking outside the
box" are recent ones. How often do managers who never take risks
themselves or think outside whatever box, urge their technical staff
to do just that? And, when a high-risk, outside-the-box project
proves fruitless, who do you suppose suffers the consequences?

of a scientific leader, the more freedom you will have to follow your own research ideas. This is a real incentive, I can assure you.

Management suggestions of an important research project or area, incidentally, need not always be bad. Michelangelo was asked by the pope to paint the Sistine Chapel. He didn't write his own proposal to an "Arts Council of Rome." Although research driven by applications is often viewed with some disdain, the desire to fulfill a real need can and has led to extremely important basic science—for example, the Nobel prize–winning invention of the transistor—and has changed the world. You can and should judge your superiors' suggested research ideas thoughtfully and on a case-by-case basis.

If you are considering a job in a commercial or government lab with the idea in mind that you will make a name for yourself and then return in style to academic life, you must be careful to determine whether your projected position and laboratory policies are consistent with your plan. If the research group you are considering works in an area that is important to the company in question but is of little basic scientific significance, you will very likely not be a viable competitor for an academic position several years down the track. You will have attended the wrong meetings, and your papers will not have been read in the academic world.

If your scientific results are going to be treated as proprietary information, i.e., are not going to be published, to protect commercial advantage, or if they are going to be hidden from the outside world as "classified data," you will not be able to achieve recognition comparable to that of many of your contemporaries. Thus, even though their scientific competence may be no greater than yours, many of your peers will have a significant advantage over you in the competition for tenured academic positions.

Apart from problems in dealing with management, one of the worst features of scientific life in many industrial and government labs is a lack of helpers. Whereas a well-funded university professor can enlist an army of students and postdocs to bring projects to fruition faster, a staff member at a research lab is lucky to have a technician and an occasional postdoc. (This is much less of a problem in the biotech industry than in companies that perform physical research, according to my sources.) There are opportunities to alleviate such a shortage, for example, by collaborating with a university research group. However, such opportunities must be aggressively pursued and are unlikely in unfavorable geographic situations. Scientists who have dreams of attacking a problem from many sides at once will not be able to fulfill them at a government or industrial lab unless they can persuade colleagues to help.

Money

In deciding what kind of scientific position to aim for, you will certainly want to consider relative pay scales. There are dramatic differences between universities and research labs in this regard. Whereas the salary distribution for government or commercial labs is a relatively narrow bell curve whose peak is in the realm of the upper–middle class, the histogram for the professoriat is much broader.* The university pay scale starts lower than in industry, and the median university salary is also lower. On the other hand, the incentives for senior scientists at a university are substantially greater than at a national or commercial lab. If as a professor you bring in substantial grant money, you are

---

* At state-funded universities, salaries are typically public information, making it possible to compile a histogram in the campus library. In some cases, publication on the Internet makes life easier. For instance, an Internet search for "faculty salaries cavalier" turns up a list compiled by *The Cavalier Daily* of faculty members and their salaries at the University of Virginia at Charlottesville. In 2008, the ratio of highest to lowest pay among biologists, chemists, and physicists on the list was about 6:1.

very valuable to your university and, not surprisingly, you reap big rewards. The ratio of highest to lowest salaries in a physics department might be 3:1 or 4:1, or more. In an industrial lab it is likely to be less than 2:1. In addition, at a university you can supplement your income by consulting and by writing textbooks on university time.

Financial priorities thus dictate the same career path as the scientific ones. Entry-level salaries are better in the research labs, and the merit pay increases they provide can keep you earning more than your university colleagues until you reach the somewhat poorly defined level of "senior scientist." After that, if you want to maximize your salary in industry or in a government lab, there is no alternative but to move into a management position. (One thing managers seem to do very well is reward themselves.) If you want a high salary while keeping a hand in research, the best alternative is a full professorship. Having established an outstanding scientific reputation working eight hours a day at a commercial or government lab, you will know what a good contract proposal looks like; you will be relatively successful at bringing in money; and so you will have a good salary, many students and postdocs, and all the good things a university has to offer.

Circumstances—economic, family, or other—may prevent you from following the optimal career trajectory. But at least I hope you will now go into the job market with a clear idea of how you would like to arrange your career and why.

ADDITIONAL READING

Browse sciencecareers.sciencemag.org, the careers website of *Science* magazine.

CHAPTER 7

Job Interviews

Succeeding in a job interview is much easier if you have an idea of what is expected of you. It is amazing how many job candidates fail because they are totally unaware of what their interviewers are looking for and what makes their interviewers nervous. Although the criteria are considerably less stringent if you are seeking a postdoctoral rather than a permanent position, the basic themes are the same: Are you a self-starter or a drone who always needs to be told what to do next? Are you a leader or a follower? Will you take an interest in your colleagues' work, or will you shut the door to your lab or office and never come out? Do you possess scientific curiosity, or do you view research as just another job? The drones, the followers, and the noninteractors, in general, need not apply.

The best preparation for a job interview, just as in the case of exams in school, is to work out in advance what questions are likely to be asked and to have answers for them. In the case of a job interview, the most important question is some variation of "What will you do here if we hire you?" A good time to prepare an answer is when you are putting your résumé together. In addition to giving you a head start on your interview preparation, if your résumé includes a persuasive paragraph or two on the research efforts you plan, it may help you land an interview in the first place.

No Dilettantes Need Apply

As is true in general, being bright, even very bright, is not enough to succeed as an employment candidate. I was recently part of a group that interviewed a young man with high grades and extremely good recommendations from one of our country's best graduate schools. Recommendations are not always trustworthy, of course. Over time, there tends to be an inflation of the praise level from any one institution since if a previous student was hired, a professor does not want to say that a subsequent candidate is any less worthy. Nevertheless, in this case we had high expectations because the recommendation came from a professor well known to members of our staff. As it turned out, the candidate, V., did appear to possess excellent analytic abilities. In his job seminar, he explained that he had developed mathematical tools that made it possible to extract useful information, in a non-prejudicial way, from an experimental technique that is widely used but was previously hard to interpret convincingly. V., a theorist, had gone into an experimental lab, perceived a difficulty in making sense of the data that were being obtained, and, by eliminating that difficulty, had made an important contribution. This is how he had won, and why he deserved, high recommendations.

The downside appeared after the formal talk. A member of the audience said he thought that V's new technique could be applied to a considerably wider class of experiments and gave some specific examples. V. appeared to be unaware of the opportunities to exploit his success and thereby not only to make himself useful to many others but also to achieve much wider recognition for his work. What is worse, he didn't seem to like the idea. In our private interview, V. explained that he did not want to be pigeonholed as an expert in one particular area. He thought that if he exploited his success, he would lose the freedom to work in other areas later. V. appeared fixed on the notion that he had the potential to contribute in so many areas of research that it would be dangerous to focus on any one of them for very long.

To his interviewers, the message was that V. is and wants to remain a dilettante. V. said that if he were hired as a postdoctoral researcher, he wouldn't want to work on a specific project or even in a specific group. He would want to spend a month or two on arrival looking around the lab for something "interesting" to work on. He said he was a "generalist." I wanted to know if V. thought he could find enough experimentalists at our lab who needed help understanding their data that he could make a career of work similar to that of his thesis. He said he preferred analyzing the errors of others to making his own mistakes in the attempt to create new knowledge at the forefront.

For all his brainpower and wonderful academic pedigree, and despite his real contributions, V.'s interview trip was a failure. It would certainly have been too risky to hire him in a permanent slot. He seemed much too immature.

It was even worrisome to imagine him as a postdoc. After two years, would V. have found something interesting enough to work on? Would he be salable for a permanent position at that point, or would we have to worry about his struggle to avoid unemployment?

The Employer's Viewpoint

It is important to understand the job interview from the perspective of the employer. He probably does not fill research positions very often. His research staff is generally not very large, and if the staff is broken down by subfield, the number of staffers with whom you might collaborate is even smaller. Therefore, offering to hire you is a big risk. Start-up funds are limited. Lab and office space is hard to come by. If you turn out to be directionless, if you are noninteractive, if you are unproductive, you will represent a huge waste of time and resources, percentage-wise. If you are one of ten staffers in related areas and you fail, then the department is only 90 percent productive at best. If it takes "only" three years before you are let go because you are not working out, realize that three years may be almost 10 percent of your colleagues' careers, a substantial fraction of their work years during which they might have been more productive had they had another colleague who stimulated them.

Given the perceived high stakes, it is not surprising that the scientists who interview you will want considerable assurance that you will make their department a more interesting place and will not just occupy space and absorb funds. Thus, it is absolutely fatal not to have given thought to your scientific direction, not to be able to articulate what you plan to do in the next two or three years and why. Under no circumstance should you indicate that you are willing to do "whatever the department wants" or, as V. said, that you will arrive without a clear direction and then will look for something "interesting" at the lab. Being collaborative is important, but having no inner compass is fatal. Your fellow scientists hope to learn from you. If you are simply going to be another pair of hands, a technician is a lot cheaper and much less of a risk. If you imply that you will sit in your office or lab waiting for inspiration to strike, there are enough other people applying for the job who will "hit the ground running" that you will simply not get an offer.

Even if you are applying for a postdoctoral job and expect to be working under the close supervision of a professional, it is still important that you express personal interests—a burning desire to know something. The lab where you work will continue to hire postdocs after you are gone. If the word gets out that postdocs do well at a particular lab, that they end up with permanent research positions at prestigious institutions, then the best PhD's will want to apply to the lab for postdoctoral slots. If, on the other hand, it seems that after two years the lab's postdocs have not accomplished much and have difficulty finding good positions, then university advisers will likely assume that

postdocs at the lab in question are not getting appropriate guidance and will steer their best students elsewhere. Thus a laboratory has a very real stake in your
success. Its future is at issue. If you publish an important paper or two during your two years, that will be
perceived as a real contribution. If you interact con-
structively with the local staff, you will have a particularly good chance of landing a permanent position
locally. Nevertheless, from the lab's perspective, your
main task as a postdoc is to do whatever it takes to be
able to land a good job in a timely fashion when your
brief tenure is up. Your task at your postdoctoral job
interview is to provide confidence that this will be the
case.

Although you should come to an interview prepared
to describe your own scientific goals, you should realize that if your inner compass appears to point in a direction totally orthogonal to your hosts', you are
unlikely to look like an ideal colleague. Thus, you can
enhance your chances for success by spending some
time on the Internet, boning up on the research interests and accomplishments of the members of the
group to which you are applying for a job. Just as *your*
publications represent your résumé, the same is true
of the scientists you will be visiting. If you understand
your interviewers' perceptions of what is important,
you will be able to tailor your description of your own

goals accordingly. In "doing your homework," you should aim to develop a description of how your research interests mesh with those of the group in which you would like to work. (If you cannot think of a reasonable formulation, you are probably applying to the wrong group.)

Incidentally, if you are interviewing for a professorial position, you can expect to be asked what courses you would like or be able to teach. If you are unprepared to answer this question, your commitment to being a good departmental citizen may come into question. This, then, is another area in which doing your homework might make a difference.

A few days after your personal interviews are done and you have gone home, staffers you visited will be trying to remember what you said in order to write up impressions of your performance. If you were able to ask intelligent and pointed questions about various staff members' work and to explain how your research will complement their own, their memories will be excellent, and it will be easy for them to write glowing reviews. If you hadn't a clue what is going on in their labs and expressed no understanding of how your work might help them achieve their goals, their memories will need refreshing, or perhaps they will be wondering whether you have the desire or the ability to make a serious contribution.

Remember How You Get to Carnegie Hall

Practicing your thesis presentation or seminar before your interview trip is absolutely vital. If you are comfortable giving your talk, your audience will feel more at ease and more willing to accept what you have to say. If you have dealt with tough questions before, being subjected to aggressive interruptions will not be as likely to make you defensive or make you want to find a hole to crawl into.

To this end, it is a good idea to practice at your home institution by giving your talk not just to your thesis adviser's group or a collection of your friends but to a wider representation of your department. Apart from helping you refine your understanding of your own accomplishments, responding to their expressions of incomprehension will make it easier for you to be quick on your feet when you are out job hunting. Every lab values staff members whose sharp questions at seminars expose the important qualifications of the science being presented. Thus, you can be almost certain that there will be an inquisitor or two in the room trying his best to make you squirm—often it will be the last young scientist to be hired, trying, consciously or not, to impress the older staffers with how valuable an asset he is. You will feel and look a lot better if you are prepared to deal with this aggression. If someone raises an issue you had not thought of, you will not find yourself cringing or spluttering, but instead responding that the point in question seems cogent and is one you will certainly be investigating in the coming months.

In succeeding chapters concerned with grant applications and developing a research program, you will read words very similar to those you have read here. The preparation you make for your job interviews should in no sense be thought of as just an exercise necessary to land a position after your PhD. Thinking about what you want to accomplish as a scientist, trying to grasp the big picture that makes your accomplishments meaningful, and learning what excites your colleagues—and why—are all vital for your success after you have won a junior position. The thinking, résumé writing, and literature searching that you do in order to succeed in your job hunt will make it much easier for you to prepare successful grant applications and to decide what research projects you will want to do. When you arrive at a new job, it is very likely that your life will switch to "fast forward." The time between your arrival and when you have to be renewed, be considered for tenure, or return to the job market will seem very short and very precious. Whatever thinking you have done in advance and written preparation you have made will lighten your burdens and may keep you out of the panic mode.

Responding to a Job Offer

In the happy event that you receive one or more job offers, in addition to selecting the one you want to accept, there may be some negotiating to do. If you are a hot property—for example, if you received some special recognition for your thesis or postdoctoral work—or if you have several offers from prestigious institutions, you may be able to negotiate a higher salary from the one where you would like to work. Generally, however, at the junior scientist level, there is little flexibility regarding salaries. On the other hand, there is considerable latitude concerning start-up funds, lab space, the assistance of technicians, and other working conditions.

Because your scientific productivity on a short time scale is going to determine your job security and the likelihood of your remaining in research, you should try to arrange to have as few distractions from research as possible and to have whatever equipment and space you will need available on your arrival. There is no harm in asking the chair of a university department that wants to hire you for a relatively light teaching load for the first year or two while you are writing proposals and setting up a lab. You should also be able to specify what equipment you will need to purchase and how much it will cost and to justify these expenses in terms of the scientific output they will bring. Do not be afraid to ask for a lot, within reason. You want the department's respect, not its love.

If you examine the science world around you, you will see that *he who spends the most money has the most influence.* I do not suggest that you spend money frivolously. I know more than one young scientist who failed after setting up a lab that looked like the cockpit of a modern jetliner but had lost track of the idea that it was also necessary to generate some meaningful results. Nevertheless, if the problems you want to solve require the use of expensive equipment, you should ask for it. You certainly do not want to arrive at your new institution and then have to sit around for months unable to begin useful scientific work.

In getting the working conditions you want, the key concept is leverage. Generally, this takes the form of job offers from competing institutions. Once you have turned down your other job opportunities and are committed to the institution whose offer you have accepted, your leverage is greatly reduced. Of course, your new boss has an interest in your success. But dividing limited departmental funds is a zero-sum game, and when you arrive as a new hire, you are at the bottom of the heap, your credibility as a scientist is marginal, and therefore you are not in a good position to win battles for money, space, working conditions, or whatever. The time to negotiate is before you have eliminated your other options.

If you can manage to get the results of your negotiations in writing, it would not hurt to do so. It is not that your superiors will be intentionally dishonest. However, having your offer, in all its glory, in black and white can be useful for refreshing people's memories if the going gets rough. This raises the question of how to get a written offer without appearing to call your new employer's honesty into question. One clever strategy is to write the offer out yourself, in the following way:

Dear Dr. Honcho:

I very much appreciate the time you spent discussing my professional opportunities at LAB-X. As I understand it, the position you are offering will include the following: [Specify the important terms here: lab space, equipment, summer salary, freedom from teaching for some time, whatever.]

Please let me know whether this list accurately reflects our conversation so that we may proceed accordingly.

SINCERELY YOURS,
DR. IMA MOVER

It is not infrequent that an institution offering you a position will want an acceptance or rejection within some time limit, so it can make a timely offer or send a rejection letter to a runner-up for the job. This may put you under considerable pressure, if other places where you have interviewed are moving too slowly. If you are not prepared to answer "yes" or "no" as a deadline approaches, you should ask for more time. If the extra time is not accorded, in deciding how to respond, you should keep in mind that *your* life and *your* happiness are paramount. If you are unwilling to let go of offer number one while waiting to hear from institution number two, it might be reasonable to accept the first offer. If the later offer is better, you can take it and apologize to the first offerers for changing your decision to accept. You will not make friends by withdrawing your acceptance, and breaking a promise is certainly not something you should do lightly or often. Nevertheless, your life comes first. If an institution plays rough by pressuring you for a decision, it should be prepared to accept the fruits of its tactics. It has probably experienced such consequences before.

Keep in mind that as a junior scientist, you are the weaker party in all your negotiations. It is not for you to make life easier for the stronger parties. In general, you will not be offered a written contract or particularly good job security. Although you should consider how your handling of a job offer will affect your long-term standing in the scientific community, you should not dismiss your own needs out of hand for the sake of a potential employer's priorities.

CHAPTER 8

Getting Funded

"While you're up, get me a grant."

You have probably already heard that if you want to succeed as a professor, you will have to bring in money. You may not have learned that in the new millennium you may even have to get a grant if you want to work in a national laboratory.

In the "good old days," prior to World War II, scientists did not apply for, nor did they receive, research grants from funding agencies. Unsurprisingly, there weren't many scientists in that era. If you were independently wealthy, or perhaps if you could persuade investors to support your work, you could build up a laboratory. Otherwise, you had to make do with what your university salary and personal resources would allow. In the latter part of the twentieth century, the realization that the products of the hard sciences can protect us from our enemies, cure our illnesses, and yield products that lighten our daily burdens revolutionized the funding of science. Government and industry learned that investing in scientific leadership is necessary for prosperity (although nowadays it is no longer clear how well that lesson is remembered).

At the same time, universities discovered the blessing of receiving government and other outside funds. Although you may think that current tuition costs are astronomical, money taken in from students does not cover a university's costs. Charitable donations take up some of the slack. But major universities would have to shrink their programs considerably were it not for millions of dollars brought in via research grants. As a science professor whose salary is considerably higher than those of your colleagues in the art history department, it is your responsibility to help support yourself and your department by winning funding from the outside. If you do not, you will find yourself persona non grata. If you are untenured, you will be asked to find employment elsewhere. If you are tenured, you will be unable to employ graduate students and postdocs, your salary will diminish relative to inflation, and your influence on departmental directions will be slight to nil.

This set of realities means that if you arrive at a university as an assistant professor, it is essential for you to win a research grant as soon as possible (though, as with many of your responsibilities as an assistant professor, getting a grant is necessary but not sufficient for your job security). Because getting funded is so important, and because the demands on your time and thought processes will be very heavy when you begin your university career, I strongly recommend that you plan and perhaps even draft your proposal before day one of your university job. The best time to think about the contents of your initial proposal is when you are preparing for your job interviews. As I explained in the last chapter, your interviewers will be very eager to know what your research plans are. Thus, at the same time as you are formulating the ideas necessary to win yourself a job, and writing the "research directions" portion of your résumé, you can be writing the basic elements of your proposal. Having done this, you will be able to begin your assistant professorship with a somewhat lighter burden. If you do not know the format of a grant proposal to one or another funding agency, ask around among the professors at your current university. My guess is that they will be pleasantly surprised at your thoughtfulness about your future and glad to help out.

In writing your proposal, it is important not only to address important research issues but also to present research plans that have a realistic chance of being completed. Major initiatives that will require numerous years of labor are inappropriate for a first proposal. If you are a full professor, with several graduate students and postdoctoral associates, and if you have a record of accomplishment that proves your ability to bring a large project to fruition, then you have a chance of acquiring funds to embark on a major effort. As a beginning assistant professor, however, you have none of the above. If your stated ambitions are too unrealistic, the referees of your grant application will certainly notice and will inform the agency that solicited their opinions that competing proposals to do incremental research have a better chance of success. If you have an important idea for a major project, you can include it in your proposal as an exploratory effort along with several short-term efforts that have a good chance of being completed. Alternatively, as I discuss in Chapter 9, you can begin your major project without seeking to have it funded, spending a few hours a week on it; a couple of years down the road, you can make it the focus of another grant proposal, when it is closer to bearing some fruit.

Research grants for beginning scientists are typically awarded for two or three years (at most five). In a grant renewal application, you will be expected to report on the progress that the funding agency's money has bought. You want to be able to demonstrate some significant results. For this reason, and considering that as an assistant professor you will be spending at least half your work hours not doing research, it is an excellent idea to include in your first grant proposal some projects that are quite far along. Knowing that you will have some real successes to trumpet in your renewal is excellent for your mental health. This does not mean you should hold back completed or nearly completed research for very long. Particularly if your work is in a hot area of research, you run the risk that a competitor will publish your results before you do. That would be bad for your mental health, to say nothing of your chances for promotion.

As in the cases of writing papers and giving talks, your grant application should be generous with references to the literature. You have very little to gain by glossing over the sources of your ideas and the accomplishments of your competitors. These very competitors are going to be asked to judge your proposal. If your application appears to ignore their efforts, they will not be shy about telling the funding agency that either you do not know the literature, and are therefore likely to waste your time and the agency's money repeating the work of others, or they will say that you are so unoriginal in your thinking that you have to try to steal ideas from your fellow scientists. Neither of these comments is likely to win you support. In preparing your proposal, you should take pains to search the literature for work of a similar nature or that is related to what you are proposing. You should discuss the significance of this work in the body of your application and carefully explain how your own research will be different or will build on it, or whatever. Flattering your competitors and referees, within reason, by taking their work seriously cannot hurt your chances and may help them considerably.

A current trend in research funding is to award grants to research groups rather than to individuals. If you are asked to participate in a group grant application, you certainly ought to do so. Being a good citizen of your department is another of the necessary but not sufficient conditions for success. In addition, if the scientists in your department are collaborative enough to want to work together toward a common goal, you should take advantage of this unusual situation if you can. Nevertheless, you should realize that if the group grant is awarded, credit for bringing in the money will not be divided equally. Unless you bring something very special to the proposal, most of the credit will go to the senior members of the group. It will be assumed, not without some justification, that the success of the application was the result of their track records in research. The fact that they have found a youngster (i.e., you) to help them succeed again is a credit to them, not to you. Thus, even if you participate in the writing of a group grant proposal, you should not fail to write one of your own. When it comes time to renew your assistant professorship or to consider you for promotion, your ability to attract funding is going to be important—and it will be the results of your individual initiatives that will bring the needed recognition.

The more idealistic among you may be reluctant to apply for grant money from Department of Defense agencies or other applications-oriented institutions like pharmaceutical companies, preferring that of such agencies as the National Science Foundation (NSF) or the National Institutes of Health (NIH), which you presume to be "untainted." My own opinion is that if the money you receive is for research that you want to do, research that you think is important, you are unwise to question the motives of the agency that grants you the funds to do it. Its motives are *its* problem, not yours. I would add that money granted to NSF or NIH by taxpayers is available for essentially the same reasons as that which is filtered through the Department of Defense. People are largely motivated to spend by fear, greed, and lust. Leaving the last of these out of consideration where science is concerned, the reason that taxpayers and their representatives are willing to allocate large sums to "pure" physics research is certainly not that taxpayers are interested in arcane theories or the results of subtle experiments. It is that they believe that supporting first-rate physics research will provide their armed forces with the best weapons to defend their interests and will provide their industry with products that will keep their country competitive in the world economy. In the realm of biology, it is largely the fear of disease that keeps the money flowing into research—hardly taxpayers' fascination with the workings of the cell.

If you have a good idea for a research project, you should submit it in the form of a grant application to as many agencies as you think might be interested in funding it, tailoring the introductory remarks to the goals of the various agencies. Your chances of winning funding from any one agency are poor enough that if you allow inappropriate scruples to stand in the way of submitting applications, you may find yourself unfunded and out of scientific research entirely.

What Your Proposal Should Say

A new grant application should persuade its judges of two main ideas: 1. that the work you propose to do is important and timely; and 2. that it is realistic to suppose that you can muster the resources to fulfill your promises. The first section of your proposal should provide the background for your ideas. You should point out what you intend to learn and how the accomplishments you hope for will fit in with or revolutionize current scientific thought or our ability to acquire important information.

In areas of research that have been popular for some time, the boilerplate quotient of the introductions to proposals is often quite high. Scientists have been promising to deliver solutions to the same important problems year in and year out. With this in mind, it is a good idea to be modest in making promises, thereby showing your awareness of the distinction between pie in the sky and what you can realistically expect to achieve. You can point out the long-term dreams that have motivated spending in your area of research without pretending that your two- or three-year contribution is going to change history. Without being unnecessarily modest, understatement is likely to win you more respect than overstatement of your possibilities.

Here is an example of what I mean, an introductory paragraph for a hypothetical proposal in my own field, the science of solid surfaces:

One of several reasons that research in surface science has been actively pursued for the past several decades is that vastly important chemical reactions, from the elimination of noxious gases in automobile exhaust to the production of petrochemicals, are catalyzed on the surfaces of appropriate powdered metals and oxides. Learning to make commercial catalysts cheaper and more efficient is thus a goal worth hundreds of millions of dollars to the world economy. Surface scientists often point to this fact, despite the common knowledge that forty-some years of surface science have not led directly to a single industrially significant, new catalyst material. The reason for this "failure" is that chemical catalysis on surfaces is a very complex affair, and even the elementary processes that together comprise a catalytic reaction, such as the dissociation and sticking of a molecule to a surface, are not very well understood. One area where surface scientists *have* made significant progress is in developing tools to determine the arrangement of atoms at a surface. As a result of this progress, the atomic arrangements of quite a variety of crystal surfaces are now known. Surface science has therefore turned to the study of elementary molecule-surface interactions. By pursuing this kind of work, for example, by studying both theoretically and experimentally how a simple molecule like H_2 interacts with a relatively simple metal crystal surface, we believe that we are taking important first steps toward under-

standing the elements of molecular chemistry on catalyst surfaces.

Notice that surface science pie in the sky has not
been ignored in this paragraph. The underlying reason
for the work to be performed is that it will ultimately
lead to inventions worth billions. However, the writer
makes clear that he does not expect the contract managers to believe that his work is going to have a direct,
and enormous, economic impact. The author wants
funding to address an important science problem
whose solution will bring us one step closer to realizing
a long-term dream.

It is important in explaining the background for
your proposal to provide credible evidence that your
objectives are realistic. Thus, you should describe your
own recent progress and explain how it motivates the
work you will do, or if you are starting in a new direction, you should describe the publications of others
and point out how they suggest new efforts. If you have
developed a new technique and plan to use it in the
proposed research, you should explain the technique
carefully enough that your referees can understand it.
Your fears that your competitors may try to steal your
methods for their own use may be reasonable. Nevertheless, if you do not explain what you plan to do in
enough detail, reviewers might find your plans hard to

take seriously. Life is full of risks. This is one you will just have to take.

Funding agencies specifically ask the referees of grant proposals to evaluate the impact of the proposed work should it be successful. It is a good idea to be helpful along these lines. You should provide an overview of the field you plan to work in and make clear how the research you will do will be important if it succeeds. It is essential to show that you understand the big picture. This means your proposal writing is actually an important scientific exercise, not merely a pedestrian attempt to extract money from the government. If you can persuade yourself and others that your work represents an important piece of a jigsaw puzzle, you will find it much more exciting and rewarding, and your colleagues will take you more seriously. Say what kind of information your work will make accessible that previously was not. Explain what mystery has been impeding intellectual progress in your area. Describe why the isolation of a certain reagent is likely to be important, or why the interpretation of a previous experiment was misleading and how it confused later work. Generally, show that you appreciate the intellectual history of your field and that your work is intended to provide new and important ideas.

In writing your proposal, remember that both the referees and the contract monitors who will be judging it are professional scientists, with a good understanding of how research works. They will know, in particular, that research projects often lead in different directions from those that were planned, that ideas that seem wonderful at the outset can lead to dead ends, and that new results can appear out of the blue that make it reasonable to abandon a planned project in favor of another. Your proposal should be coherent and make sense at the time it is written. If a year later you think it reasonable to adopt an alternative approach or start on a new project, you needn't fear for your proposal's renewal, provided that you have or are very close to having obtained significant results when renewal time arrives. A grant does not bind you to follow a path that is shown to be a false one in the course of your work. The success of your application depends on your demonstrating that you have picked a good problem at time zero and of your renewal on the salability of your product after two or three years. This means that in writing your grant application, you should not try to cover all bases by writing down every conceivable approach to your problem. Make a good case for one or two projects and mentally reserve the right to do something different if those do not work out.

The preparation necessary to win research funding, in sum, is very similar to that required to succeed in a job interview and to establish an effective research program. At some stage in your life, when you are managing several research grants and graduate students and postdocs, it may be reasonable to view the writing of a grant application as a time-consuming chore. For you, however, a beginning scientist, the exercise of preparing a proposal is an integral part of what you must do to make the transition from someone who is technically able and somewhat knowledgeable to a real member of the scientific community.

CHAPTER 9

Establishing a
Research Program

I wish I could tell you how to go about winning a Nobel
prize. (I wish I could tell myself!) However, my goal in
this chapter is considerably more modest. I want to
help you see how the research program you establish
will affect your chances not only of producing important science but of staying in science at all.

To succeed, you will have to make a rather coldblooded analysis of your capabilities. This means planning not just scientifically exciting projects but ones
you can complete in good time. You need to consider
how your present activities will affect your long-term
interests. This may lead you to broaden your efforts

well beyond the field you were hired to work in. On the other hand, you should recognize when your experience gives you an advantage relative to your competitors—a special perspective based on your work in another field, or an unusual technical capability—and choose projects that exploit your advantage.

Although it is a good idea to build on your experience, whether by using novel techniques you have developed, complex ones that you have mastered, special reagents you have purified, or organisms that you have isolated, you will greatly improve your chances for long-term productivity and survival in research if you can teach yourself to be problem- rather than technique-oriented. Problem-orientation means keeping clearly in mind the scientific problems you want to solve and working toward their solution even if it means learning or developing a new technique from time to time. You want to be more than simply the master of a particular technique, uninterested in any scientific issue to which it is not applicable. If you operate in the technique-oriented mode, you are unlikely to be a scientific leader for long, and your freedom to pursue personal research interests will probably not last. Being problem-oriented does not mean you need to master *every* technique necessary to solve a problem of interest—often it will make more sense to take on a collaborator than to learn yet another method. What it does mean is that you will be primarily a *scientific* leader and only secondarily a *technical* one.

Some fields of research are riskier than others. For example, if you work in an area sufficiently developed that there is just one "big problem" to solve, the chances that you will be the one to solve it may be rather slim. Starting your career off in an area where your contributions have a better chance of gaining recognition would seem more sensible, if somewhat less exciting.

Timing Is Everything

Timing is one of the most important issues in establishing your research direction. A problem that will take two years to finish must not be the main focus of your activities if you are a postdoc and will be looking for a permanent position in a year and a half. If your postdoctoral adviser suggests that you work on a major, long-term project, you should at the very least ask for an estimate of what you will have to show for your efforts by the time your job hunt is to begin. You might also ask whether you will continue to receive financial support if your results are still several months off when your postdoctoral term is due to end. If you hold a two-year position and your adviser cannot persuade you that your project has a reasonable chance of yielding publishable, noteworthy output within 18 months, say, respectfully, that you need to start on some short-term research efforts first, or perhaps simultaneously. If your adviser insists that you devote yourself wholly to the long-term endeavor, remember that ultimately *you* are responsible for your success or failure as a scientist. If your adviser (especially your young adviser) places his or her interests above your own, do not be too surprised. Seek a different group to work in, one that offers you a more realistic opportunity to produce short-term, publishable output.

In looking for an alternative research group, do not whine about adviser number one to prospective adviser number two. Your goal in interviewing for a new opportunity is to persuade the new group leader that you are mature enough to understand what is necessary to launch your career. Without complaining, you can make clear that although your initial adviser's project is one you would have liked to pursue, you fear that you are not going to be around when the important results are obtained and published, that you will get little credit for your contributions, and that you want to avoid living on an unemployment check two years hence.

Timeliness Versus Importance

Apropos "coldblooded analysis," the idea that the importance of a project justifies a long-term effort is worth a critical look. Experience teaches that, important or not, a research endeavor becomes *timely* only once it can be approached with suitable technical infrastructure. Before then, a proposed long-term effort is likely to translate into fruitless weeks, months, or even years of struggling to make headway with inadequate tools.

Because beating your head against a wall is neither satisfying nor productive, you should be wary of embarking on long-term efforts, whether formulated by yourself or suggested by a mentor or collaborator. It may make better sense to put off work on that important problem until new techniques have been developed—perhaps by you, perhaps by somebody else—than pushing ahead, on the assumption that brute force will eventually lead to success.

Apart from whether you will be able to obtain significant results before your return to the job market or your consideration by a tenure committee, a serious peril of the brute-force approach is that a competitor will develop a labor-saving new technique and race to the goal while you are still struggling.

Technique- Versus Problem-Orientation

Most young scientists emerge from graduate school having learned a set of technical skills. Many are tempted to try to build a research program around them. This frequently leads to an unfortunate mode of thinking about what to do next, which I call, with apologies to Luigi Pirandello, *Six Techniques in Search of a Problem.* The institutions that hire young scientists often reinforce the technique-oriented approach to research planning by looking for new PhD's or postdocs who have worked with a particular instrument—for example, at a synchrotron radiation facility—or who have experience with a hot new technique—such as scanning tunneling microscopy or transgenic organisms. If a new hire swallows the idea that he is to be "the man at the synchrotron," and particularly if he feels that he must reject any project that does not involve synchrotron radiation, he is likely to have little impact on the world of science, with corresponding consequences to his career.

When a remarkable new instrument, such as the laser, or a technique, like nuclear magnetic resonance spectrometry, becomes available, it is often profitable to ask how its capabilities can be applied to solving outstanding problems. Few scientists, however, are able to make a long-term success of applying their favorite technique to one problem after another. Eventually the well runs dry. It is the researchers who focus on a significant problem and are willing to bring to it whatever resources are necessary who give the most absorbing talks, write the most significant papers, and win grant support most easily. I strongly recommend that you try to teach yourself to be problem-oriented, to plan your research projects so that they address important scientific issues regardless of what techniques you and your coworkers will need to use.

The people who hired you because of a certain technical expertise may be somewhat to very disappointed when you first announce that you will not be spending *all* your time working with the synchrotron, scanning tunneling microscope, or whatever. On the other hand, they will not be pleased, some years later, if you have become obsolete along with your particular technique. If and when you decide you need to branch out or move away from your initial technical role, you must make certain to fulfill your commitments to ongoing projects. Assuming that you do this gracefully, your group's disappointment at your change in technical focus will be tempered as your broadened effort leads you to the solution of an important science problem, enables you to win new research funding, and maintains or enhances your standing in the research community.

Strategic Thinking

There are several strategies for establishing a record of accomplishment that will help make you more salable or will enhance your chances of winning promotion to a continuing scientific job. The most obvious is to aim at an important long-term goal by planning your work as a sequence of short-term projects. Each of the latter should yield an identifiable and publishable milestone (a "publon"; see Chapter 5). Your papers and oral presentations can then begin by identifying you and your work with an exciting research area, while the new kernel of knowledge that you describe will give confidence that you are a person who completes projects and who will be a credit to the department that hires or keeps you.

Planning and publishing the results of short-term projects minimizes your chances of being scooped. No matter how clever you are, and particularly if you choose to work in a fashionable research area, you will have some very clever competitors. Packaging your ideas in publishable bundles and getting them out into the literature is important if you are to get credit (to "establish priority") for your work. Apart from enhancing your personal scientific reputation, this is important to the people who pay for your research and want recognition for that.

Each time you lengthen your publication list by publishing the results of a short-term project, you lower your risk factor in a potential employer's eyes. A proven producer is always preferred to a pig in a poke, and a substantial publication list is the best evidence that you have been and will be productive. Although professionals rightly scorn colleagues whose publication list is padded by repeated articles on the same work, you win no brownie points for writing long, multifaceted papers (cf. Chapter 5). Each time you publish the results of one of your short-term efforts, you advertise your productivity and that of the institution you work in to your fellow scientists, your contract managers, and your potential future employers. You also perform an estimable service to the research community because the timely introduction of new ideas speeds up the development of a field and prevents duplication of effort. There is always an opportunity to write a comprehensive review when several small projects add up to a major accomplishment or discovery.

Incidentally, publishing more papers rather than fewer will help you in several ways with the bean counters among those who judge you. They will not only look at the number of papers you have published but will also consult a citations database (e.g., the ISI Web of Science[SM]) to see how many pages of citations your papers have garnered. If you have published twice as many articles, this "objective measure" of their impact will be roughly twice as great. You may find this idea crass. I do. But it is safe to assume that there will be bean counters among those who determine your future, and it certainly does you no harm to please them.

Another important strategy for establishing a successful scientific career is to work on more than one project at a time. This has several advantages: It means that when you temporarily run out of ideas related to project A, you need not waste the rest of the day, week, or month but can simply turn to project B. When a project has been completed, you do not have to spend entire days wondering what to do next but rather can budget some time to push ahead on another, one hopes publishable, piece of science.

Working on more than one project is the only way a young (or any!) scientist should undertake an inherently long-term project. I spent ten years (!!) writing a computer program to model the energetics of atoms and molecules on metal crystal surfaces. Although I was able to publish several pieces of technical progress along the way (e.g., mathematical tricks that made portions of the computation more efficient), the really significant science output could only be produced when the computer code was substantially complete. I survived this project scientifically by establishing collaborations in which the tools required to generate results were either completely or almost completely developed. By devoting about 50 percent of my time to short-term projects using these tools, I maintained a publication record—several new papers a year—adequate to persuade my peers and my employer that I was not brain-dead.

I do not, by the way, recommend ten-year projects as a good idea for young scientists. I waited until I had established a strong scientific reputation before risking it. But even if you want to carry out a three-year project, having something else going on is highly recommended.

Working on two or three projects simultaneously has at least two other advantages. One is that it forces you to be broader than otherwise. There is a strong tendency to become narrower and deeper as you progress scientifically, particularly if you work in an industrial or government laboratory. At a university, teaching requirements counteract this tendency. Without at all wanting to argue that you should strive to be broad and shallow or that you should spread yourself so thin that you are unable to make progress in any area, I suggest that by having your fingers in several pies, you are more likely to prosper scientifically. As one area loses its scientific appeal, another with which you are already familiar may increase in importance. The clever ideas you learn or develop in one area may be applicable in another. This can be an extraordinarily efficient way to make progress.

The second advantage of having more than one project underway is that it will lessen the impact on your career should you be scooped. This is something to worry about if you have chosen to work in a hot area.

Establishing a Name for Yourself

It is particularly important that a young researcher establish an identity in the community. Collaborating with other scientists is certainly an effective way to build up a publication record. However, except under special circumstances—for example, if you bring a unique and identifiable skill to the collaboration—most of the credit for the papers you write will go to the senior partner. Instead of your work's being referred to as "Young Postdoc, et al." it will be the paper published by "Honcho's group." This is independent of the fact that your name came first on the paper.

For this reason, it is important for you to start thinking up, working on, and publishing the results of projects where you are the sole author or perhaps the only theorist in collaboration with an experimental group. In the latter case, it is not enough just to act as the house theorist, the data analyst who performed regressions on demand. You must perceptibly contribute new ideas—ones that your experimental colleagues would be unlikely to have produced on their own.

Risky Business

Although working in a hot area is exciting—major meetings are mob scenes, the scent of a prize is in the air—it is a risky business. Before moving into a fashionable field, you must ask yourself whether you have a realistic chance of emerging from the mob as someone who has made an important advance. If the problem is solved and this hot area is the only one you know well, how long will it take you to establish yourself in another one? Are your ideas sufficiently different from others' that you can hope to beat the competition to the answer?

A less risky course is to try to lead rather than follow fashion. One way is to think how a recent technical advance may have made a problem ripe for solution that had previously been untimely and therefore pushed to the back burner. Another is to make the needed technical advance yourself. That may require hard work. But in compensation, you will likely not have to race to outdo competitors; few will want to invest the labor. If in the end you make a distinct advance in the technical state of the art, you will deserve, and win, considerable recognition.

Aside from working hard, you can reduce the risk inherent in undertaking a major project by making sure that enough money is spent on it. After a research department or funding agency has invested heavily in your goals, it has a real stake in your success. It is correspondingly reluctant to admit that your project is going awry.

No one ever got ahead in science by saving money. In my own area of research, for example, great algorithmic advances have made it possible to compute the properties of solids in a fraction of the time that was previously required. Does this mean people are requesting smaller computer budgets? Not on your life! They have scaled up the size of the problems they propose to solve. They are asking for bigger computers than currently available and for more computer time.

Ambition is rewarded in scientific life. Lack of it leads to the exit. Let your management worry about pinching pennies. That is not your job. Let the people who pay the bills know you are scientifically alive not only by publishing exciting results but also by keeping up your requests for support.

CHAPTER 10

A Survival Checklist

Your surgeon should use one but may not.
Your airline pilot, thankfully, has no choice.

You have a lot at stake in your quest for a permanent research position, the investment of "the best years of your life," eight or nine of them, in science-oriented higher education, and likely several more in postdoctoral research. Referring to a checklist may help you stay on the track to success.* Here is what it should say:

---

* Atul Gawande, *The Checklist Manifesto: How to Get Things Right* (New York: Metropolitan Books, 2009).

1. Put yourself in the shoes of your audience: In whatever aspect of your scientific life—deciding how to spend your time at work, preparing a seminar, or writing and editing a manuscript—step outside yourself to imagine how your department, management, listeners, or readers will respond to your effort. To win a permanent research position is to seal a contract with the scientific community. That will not happen unless both sides are satisfied with the terms. You would not accept a job offer without guarantees of enough time to conduct your research, enough funding to get started, freedom to "get a life" outside your work, and an adequate paycheck. Now ask yourself what terms your paymaster(s) and scientific audience might require. To begin, ask the number-one question: "Would I recommend hiring a candidate who has competently taken data with high-tech instruments or learned to run a sophisticated computer program but not produced a publication?" The answer, "No," is invaluable guidance. It is a reminder that finishing projects, writing them up, and sending them off to journals are prerequisites for winning the job of your dreams.

Ask yourself next, "How would I react to a poorly prepared interview talk? If the slides were cluttered and confusing, if the arguments were unconvincing, or worse, would I be excited about hiring the speaker?" The obvious answer is ample impetus to make your oral presentations, all of them, engaging, informative, and persuasive. (Return to Chapter 4 for details.)

Last, ask how you would react to a badly written research statement or paper on a candidate's publication list. Not well? Again putting yourself in the shoes of a potential employer, you will realize that becoming a merciless editor of your own writing is an excellent investment of your time. (Return to Chapter 5 for a review of specifics.)

In short, to *get* the research job of your dreams, you must learn to *give* what your audience wants. Putting yourself in their shoes is the best way to understand what that is.

2. *Get your priorities straight!* Should an opportunity arise to embark on a new activity, do not say "yes" before considering a key question: "What is my job?" For a common example, imagine that several months into a two-year postdoctoral stint, while striving to complete your first research project, you notice an announcement, or your research adviser tells you of a competition for a grant. Should you compete? Generally, no. Your answer to "What is my job?" should be, "First and foremost, to complete my research project," including writing it up and submitting it for publication. It is *not* to bring in money. That is your adviser's or manager's responsibility. Once far enough

along in your postdoctoral sojourn that you have submitted a paper for publication (and more than one
would be better), and confident that you have a compelling story to tell in your job hunt, *then* you might
consider spending time preparing a grant proposal.
Otherwise, for a postdoc, grant-writing is a diver-
sion. Single-mindedness is a more likely prescription
for success.

Here is another example: Suppose, as a junior faculty
member or starting lab scientist, you are asked to serve
on a committee of a national scientific society. Should
you say yes? The answer depends on anticipated work
load. Decline if the job will be so burdensome that it
stands in the way of your producing new science at an
acceptable rate. You may win esteem by serving on a
national committee and may well build a network of
influential colleagues there, but at crunch time (e.g.,
when you come up for tenure), you will be judged by
your scientific output before anything else. Once you
have won permanent employment in the research
community, you can serve on all the committees you
want. Not before.

3. *Learn when to say no:* Can you survive as a cheerful
scientist without being somewhat selfish? Likely not.
If you are perceived as someone incapable of saying no
to committee work, or to becoming associate editor of
a journal, or to being the lead investigator—the one re-

sponsible for collating all the many contributions—on one joint project after another, you are likely to join the many scientists who are perpetually under stress and who often seem irritable or angry. Your goal, whether in a university department or a research lab, is to win respect for your scientific product, not love for taking on whatever extra job comes your way. Once you are established and correspondingly experienced, you will have the freedom and ability to multitask. Even then, however, declining more than a little extraneous work is likely to make you (and your family) happier.

4. Be thoughtful about networking opportunities: Beyond being scientifically productive, is there a surer way to the job of your dreams than through connections? How do you become a member of the old-boy or old-girl network? Not by learning a secret handshake, but by taking advantage of opportunities to make yourself known.

Begin at your desk. Have you read a stimulating paper related to your work? Has it raised compelling questions? Engage the author in an email dialogue. When you start looking for a job, he or she might recall your thoughtful queries, or your critique, and be willing to help.

At the lab where you now work, budget time to learn what people beyond your research mentor's labs are doing. Attend their seminars. Engage them in dialogue.

Are you about to attend a conference? Read the abstracts. Pick out the talks most relevant to your interests, then download and peruse their authors' recent papers. Go to the conference with questions. Meet selected authors after their sessions end.

Will you be traveling? Is there a lab near your destination where you might like to work one day? See if you can arrange a lab visit while you are in the neighborhood. If you are still a student, apply to spend a week or a month there during the summer. In the mode of "putting yourself in their shoes," think how much easier it is for an employer to hire someone he or she has met and sized up, compared with another who has come for a brief interview visit—and whose recommendation letters may be inflated. Thus, aim to be the person your hoped-for employer already knows.

After giving a career day lecture at a midwestern state university several years ago, I was asked whether the ideas I had presented wouldn't lose their advantage if everyone adopted them. "That's true," I replied, "in the sense of the theory of the efficient marketplace—but I'm not holding my breath." This chapter's checklist largely amounts to common-sense ideas. But common sense is in shorter supply than you might imagine, and the market for permanent positions in research is correspondingly far from efficient. Thus, mind the checklist to stay on track; many others won't.

Afterthoughts

Experience is the best teacher (but only when the experience isn't fatal).

The tacit premise of this book is that behaviors appropriate to launching a scientific career can be learned. Many of my colleagues doubt this, throw up their hands, and propound the Darwinian approach. They say that scientific maturity comes with experience and cannot be taught. The fittest students will survive. The rest will not, according to the law of the science jungle. As I mentioned at the outset, adopting this fatalistic, laissez-faire viewpoint does have the advantage that busy professors need not spend time trying to teach their students science survival strategies. On the other hand, if they are wrong, then they are guilty of avoiding an important responsibility.

I take a behaviorist viewpoint. Although the inner feelings and thoughts that go along with scientific maturity may be real and may only come with experience, what is needed to make the transition from graduate student to professional researcher is to learn certain behaviors. It is not important whether a student prepares an adequate introduction to a seminar because my book suggests it would be a good idea, rather than because of a deep inner conviction based on experience. What *is* important is whether the seminar ends up stimulating and enlightening listeners. Arguments over the possibility of teaching students to be mature should not stand in the way of teaching the skills involved in giving good talks, writing excellent papers, succeeding in job interviews, and so forth. They are not all that hard to learn, and the underlying ideas do not tax one's intellectual powers greatly. It should be obvious that the problem with waiting for experience to dictate appropriate behaviors is that one is very likely to fail as a result of the bad experiences that are supposed to produce the appropriate feelings. *It is far better to learn from the bad experiences of others than from your own.*

The result I have hoped for in writing this book is that you will become more reflective about your career and act in a way that is appropriate to being successful and productive. If you stop to think about whether that talk you have been working on is well organized, whether the paper you are writing is one you will be proud of in five years, or whether the research program you have developed is appropriate to your station in scientific life, I will have succeeded. No matter how well you do in these regards, you will certainly still experience difficult times, have regrets about some of your choices, and possibly fail anyway. Nevertheless, your chances for having a scientific career will be greatly improved.

I wish you every success!

Readers' Suggestions Are Welcome

My view of the world of science is inevitably framed
by my own experiences and those of my colleagues.
You can help subsequent editions of this book reflect
a broader view of what it takes to establish a scientific
career. Send anecdotes, suggestions, criticisms, and
comments to me, care of:

Basic Books
387 Park Avenue South
New York, NY 10016

Thank you in advance for your help!